T0203985

Plant Pathogenesis *and* Disease Control

Hachiro Oku
Laboratory of Plant Pathology
and Genetics Engineering
College of Agriculture
Okayama University
Okayama, Japan

CRC Press
Taylor & Francis Group
Boca Raton London New York

CRC Press is an imprint of the
Taylor & Francis Group, an **informa** business

Published 1994 by CRC Press
Taylor & Francis Group
6000 Broken Sound Parkway NW, Suite 300
Boca Raton, FL 33487-2742

First issued in paperback 2019

ISBN 13: 978-0-367-44969-8 (pbk)
ISBN 13: 978-0-87371-727-4 (hbk)

Visit the Taylor & Francis Web site at
http://www.taylorandfrancis.com

and the CRC Press Web site at
http://www.crcpress.com

Library of Congress Cataloging-in-Publication Data

Oku, Hachiro, 1927-
Plant pathogenesis and disease control / Hachiro Oku.
p. cm.
Includes bibliographical references and index.
ISBN 0-87371-727-9
1. Phytopathogenic microorganisms—Control. 2. Plants—Disease and pest resistance. 3. Phytopathogenic microorganisms. 4. Plant diseases. I. Title.
SB731.027 1993
632'.3—dc 20 **93-21626**
CIP

Library of Congress Card Number 93-21626

Preface

The environmental pollution caused by pesticides used for pest control of cultivated plants is a worldwide problem. Now we must change the concept of "pesticides" to "pest control agents", especially in plant disease control. Noncidal disease control agents might have less risks to other living beings than plant pathogens, and hence, to the environment.

In the book, *Development of New Pesticides*, published by Nankodo, Tokyo (1965), I wrote a chapter called "Disease Control Mechanisms of Pesticides", in which I described the usual developing procedure employed in the screening of substances that have antifungal activity among the compounds synthesized at random. If some compound has antifungal activity, then its derivatives are synthesized and tested for antifungal activity. However, this method has come to a dead end. To break this impasse, it is most important to know the mechanism of the outbreak of disease in order to prevent it. Theoretically, compounds that inhibit the penetration of pathogens into plants, that inhibit the biosynthesis of pathotoxins, or that activate the resistance response of plants against disease might be more promising for future pesticides because they would not cause pesticide pollution.

At the time that chapter was written, I was a research scientist in the Pesticides Research Laboratories of a chemical industrial company and was always criticizing the routine developing process of fungicides as described above. This screening process can be accomplished by laboratory technicians and does not require plant pathologists. A chemist who was my colleague read my chapter and said to me, "Dr. Oku, you are dreaming!".

However, 26 years later, we have many effective noncidal disease control agents belonging to the above categories and "my ancient dream" has become a reality. These agents contribute much to plant disease control without environmental pollution.

Since 1968, when I moved to a university from the industrial company, I have devoted all of my academic life to studying the mechanisms on the pathogenicity of fungal plant pathogens, which might present the fundamentals to develop control measures for plant diseases without environmental pollution by fungicides.

This book is written from such a viewpoint and I sincerely hope it will be a milestone for the development of harmless control measures for plant diseases.

Hachiro Oku
Okayama, Japan

The Author

Hachiro Oku, D.Sci., is a Professor Emeritus, Laboratory of Plant Pathology and Genetic Engineering, College of Agriculture, Okayama University, Okayama, Japan.

Dr. Oku graduated from the Gifu College of Pharmacy in 1948. In 1951 he graduated with a B.S. degree in Plant Pathology from Kyoto University, Kyoto, Japan. He earned his D.Sci. degree from this same institution in 1962.

Dr. Oku is a trustee for the Phytopathological Society of Japan, the Pesticide Science Society of Japan, and the Japanese Society of Plant Physiologists. He is also a member of the American Phytopathological Society.

He has received awards from the Phytopathological Society of Japan (1977) and the Association of Agricultural Science Societies of Japan (1992). In 1992 he also received the Yomiuri Press Agricultural Award.

Dr. Oku has published over 140 articles in scientific journals and is the author of several volumes about plant pathology. His current research interests include mechanisms of pathogenicity of fungal plant pathogens, and the development of control measures for these diseases without environmental pollution.

Contents

Acknowledgments

The author acknowledges with pleasure Professor T. Shiraishi and Drs. T. Yamada and Y. Ichinose of our laboratory for their generous assistance and helpful suggestions.

The author also thanks Mr. T. Kato, graduate student of our laboratory, for preparation of the tables and figures.

INTRODUCTION

Basic Principles for Plant Disease Control

I. FUNGICIDES AND BACTERICIDES MAY HAVE RISKS FOR OTHER LIVING BEINGS

If our only purpose is to protect crop plants from diseases, the routine application of fungicides or bactericides might be most efficient. Pathogens reaching their host plants may be killed by the biocides residing on the plants and eliminate the incidence of disease. Even if some individual pathogens escape from such biocides, the next application of the biocide may kill the pathogen.

Each country and its districts have respective traditionally important crops according to their own environmental conditions. These important crops are invaded by several destructive pathogens. The famine of Ireland caused by the potato late blight disease is one famous example. Ireland lost roughly one third of its population within 6 years in the mid-19th century.[1] The most devastating crop disease in Japan has been the rice blast disease. Ancient records suggest that many famines — in 1695, 1783, 1833–1837, etc. — seemed to be due to this disease; and many people died of starvation.[2] According to statistics compiled by the Japanese government,[3] about 80% of the yield reduction was due to the blast disease during the years when the damage of rice by disease was severe.

After World War II, organic mercurial compounds which were developed in Germany as a seed disinfectant, were found to be very effective against the rice blast disease, and these compounds eradicated this disease from rice cultivation in Japan. At that time, organic mercurials had been used only from the viewpoint of disease control without considering the side effects; however, in 1974, the use of these mercuric compounds as pesticides was prohibited by the Japanese government for sanitary reasons.

In general, compounds which kill the pathogenic microorganisms (fungicides or bactericides) may be more or less toxic to the other organisms, because of the small difference between them at the cellular level — the smallest unit of life — both in function and structure. The injurious effects of biocides to organisms other than pests are the causes of pesticide pollution such as sanitary problems, phytotoxicity, and injurious effects on the natural ecosystem.

II. THE DIFFERENCE BETWEEN NATURAL AND AGRICULTURAL ECOSYSTEMS

If we could ensure the quantity and quality of agricultural products without the use of agrochemicals, nothing would be better than that. It is possible to cultivate crops without the agrochemicals when environmental conditions are unsuitable for the growth and multiplication of pests; however, in cases where conditions are favorable for the outbreak of pests, a great famine (just as in ancient times) may occur if we do not use agrochemicals. Also we would like to consider why cultivated plants are easily destroyed by pests compared with wild plants.

It is believed that life was generated in our planet about 3000 million years ago, and living organisms now in existence have evolved throughout this long period of time. According to the theory of natural selection which sustains the theory of evolution, the direction of the mutation of organisms is random; the mutants which did not adjust to the environmental condition or were defeated by the struggle for existence have become extinct. On the contrary, organisms that adapted to the environment and acquired rapid reproduction mechanisms still exist, and have built up the present ecosystem. The different biophase in each geographic region may be due to the differences in environmental conditions to which organisms have adapted. Thus, the stable interactions during those periods of time built up the natural ecosystem with these organisms.

The members of a natural ecosystem utilize the other members as food to sustain their individual life and species. However, mankind has developed methods to cultivate exclusively plants and animals suitable for food. For this purpose, mankind has destroyed the natural ecosystem and built up an artificial one. This is the agricultural ecosystem. Although the growth of natural foods is now popular, the only actual natural foods in the biological sense are wild animals, grasses, and sea or river fish. Cultivated plants and domesticated animals are not natural foods. That is, man has to realize that agriculture is created in compensation for destruction of the natural ecosystem. The properties of destruction, however, differ from that of the other industries. Agriculture dealing with living beings cannot neglect natural conditions. For example, we cannot grow tropical crops in the frigid zone. To do so would result in a great loss of energy. Likewise, we should understand the rule of how the natural ecosystem evolved and how agricultural practices should harmonize with it. By doing this, the outbreak of pests might be diminished extensively.

III. DECLINE OF DEFENSE FUNCTION IN CULTIVATED PLANTS

What are the differences in diseases of plants in the natural ecosystem from cultivated ones?

Each member of a natural ecosystem maintains stable interactions with other members as described above. This relationship is also maintained between higher plants and their pathogenic microorganisms. This might be due to the mutual acquisition of resistance mechanisms during the long period of coevolution of both organisms, and it is seldom that any plants are exterminated by diseases under the natural ecosystem. In other words, the higher plants in a natural ecosystem have accumulated genes for resistance against the invading microorganisms. This is the reason why wild plants have been used as sources of resistance genes for breeding programs of disease resistant cultivars of cultivated plants.

The fact that pathogens newly imported from a foreign area sometimes cause devastating problems might be due to the absence of genes resistant to the new pathogens, because of the lack of the coevolutional process.

According to the review by Takai,[4] the Dutch elm disease was first found in The Netherlands after World War I in 1918. The pathogen, *Ceratocystis ulmi* is transmitted by four species of elm bark beetles, genus *Scolytus*, in Europe. The outbreak of this disease in the United States was first found in 1930, due to the import of elm logs from Europe to Cleveland, OH. The vector beetles (*Scolytus* sp.) persisted in the imported logs, multiplied, and caused disease to native elm trees in the United States. After that the native elm bark beetle in the United States, *Hylurgopinus rufipes*, also became the vector for this pathogen and expanded the area of this disease. The first outbreak of the disease in Canada was also due to the transport of elm wood from England in 1944. The disease spread very rapidly on the North American continent, and elm trees in the United States and Canada have almost been wiped out during this century.

In spite of the severity in Europe and North America, elm trees on the Asian continent and Japan are resistant to this disease. According to Smalley,[4a] this disease can be found on mainland China, but causes no serious problems. From these facts, many scientists consider that this disease originated from Asia. The breeding program to grow elms resistant to this disease is now going on by crossing *Ulmus haponica* (a species having good shape) with *U. pumila* (a species with high resistance).

The pine wilt disease is the most devastating disease of forest trees ever experienced in Japan, and has called national attention not only from environmental but also from cultural viewpoints because pine trees are deeply involved in traditional Japanese culture.

This disease has somewhat similar features to the Dutch elm disease. The pathogen of this disease, the pinewood nematode *Bursaphelenchus xylophilus*,[5,6] is transmitted by a cerambicyde beetle, *Monochamus alternatus;* and many scientists believe that this disease was imported from the United States.[7,8] That is, according to ancient records, a disease with the same symptoms as this disease was first found in 1905 at Nagasaki, the site of the most famous seaport in Kyushu Island. The first record on the outbreak of this disease in Honshu Island was in 1914 around the Aioi seaport in Hyogo Prefecture. The fact that the disease outbreak began around the port strongly suggests that the disease was imported from a foreign country. When Mamiya[8] identified the sample sent by Dropkin[9] in the United States as the pinewood nematode, the survey of this nematode was conducted actively; as a result, this nematode has been found in more than 30 states in the United States. The nematode seems to have existed in the United States since ancient times. Because the native pine species in the United States already have acquired resistance to this nematode, this pathogen does not cause as serious a problem as it does in Japan. The main pine species destroyed by this nematode in the United States are imported ones such as *Pinus sylvestris, P. thunbergii, P. densiflora,* and so on.

The facts described above indicate that plant pathogenic microorganisms keep stable interactions with their host plants under the natural ecosystem in each respective environment. The stable interaction may be destroyed by artificial factors added into the natural ecosystem.

In contrast, since crop plants are grown in an artificial environment, it is likely that there is no stable interaction between crop plants and pathogenic microorganisms. That is, cultivated plants are more susceptible to diseases.

According to the Old Testament, people in ancient times suffered much from diseases and insects that attacked cultivated plants. The unstable supply of food crops was mainly due to the diseases and the injurious insects at that time. Theophrastus[10] (370–288 BC) described that cultivated plants were more susceptible to diseases than wild ones. Thus cultivated plants have to be protected artificially from diseases and injurious insects.

The important crop plants for mankind such as cereals, potato, and rice have been improved from year to year everywhere in the world by breeding to fulfill the needs of mankind. The direction of breeding is mainly focused on increasing yield, nutritive value, and quality; and many cultivated plants at the present time are far different from the original wild species. For example, turnips have developed extremely big roots, and apple fruits are several times larger than those of the wild varieties.

These improved cultivated plants each have destructive pathogens. Breeding of the resistant varieties or cultivars has also been conducted for a long time. However, pathogens to such improved plants usually become pathogenic to the newly developed resistant cultivars within a short time. That is, new races appear after the development of new cultivars. Many races have been found in rust fungi, powdery mildew fungi on cereals, late blight fungus on potato, blast fungus on rice, and so on. The race-cultivar specificity has been explained with the gene-for-gene theory by Flor,[11] and no exceptional example has been found at present. This fact suggests that a new race might acquire new gene(s) for virulence or lose gene(s) for avirulence against new genes for resistance during the adaptation of pathogenicity to the new cultivars.

In the meantime, the improved cultivated plants sometimes return to their original wild states. For example, the big root of the turnip or carrot suddenly reverts to a slender, hard, and fibrous root. This is not a disease of the turnip or carrot, but the plant has no more commercial value for consumers; and we call this phenomenon the relative disease in contrast to the real scientific disease, absolute disease. Thus, cultivated plants are in some respects artificial plants; and furthermore these artificial plants are exclusively grown in the artificial environment of a definite area. Therefore, it is highly probable that these plants are to be the target for many plant pathogens, insects, and other organisms. Once these plants are attacked by these organisms, pests may multiply and spread very rapidly from one plant to another. Under these circumstances resistant cultivars developed by breeding may be attacked by new races of pathogens within several years. Therefore, we cannot maintain the amount and quality of food we need without the artificial protection of agricultural crops. Recent procedures for breeding resistant cultivars to disease have improved greatly, but it is difficult to stop the appearance of new races of pathogens only by breeding. It is also impossible to suppress completely the outbreak of crop diseases by devices of the cultivation procedure. Further, many accidents occurred in ancient times by eating diseased crops and moldy foods. Dietary levels of aflatoxin B1 (which is produced by *Aspergillus flavus*) as low as 1.5 and 0.5 ppb can cause liver cancer in rat and rainbow trout, respectively.[12]

Thus, we come to the conclusion that it is impossible to supply safe and adequate quantities of food materials without protection of agricultural crops by agrochemicals, especially when climatic conditions are suitable for the development of pests.

IV. UNDERSTANDING OF PATHOGENICITY, THE BASIS FOR PLANT DISEASE CONTROL

A majority of the protection of agricultural crops against pests should be dependent on agrochemicals at the present time, because we have no other procedure for supplying safe and adequate quantities of food materials to the increasing world population.

Table 1. Development of Disease Control Agents

Year	Chemicals	Effective to	Country developed
1705	Mercury chloride	Wood decay	Europe
1716	Tar, alcohol	Wood decay and insects	U.S.
1744	Copper sulfate	Seed disinfectation	France
1851	Lime-sulfur	Rust, powdery mildew	France
1883	Bordeaux mixture	Diseases of grapevine	France
1912	Organomercury compounds	Seed disinfection	Germany
1918	Chlorpicrin	Fumigation	Germany
1930s	Dithiocarbamates	Many diseases	U.S.
1960s	Blasticidin S kasugamycin, rabcide, organophosphorous compounds	Rice blast	Japan
	Benomyl	Many diseases	U.S.
1970s	Oryzemate, isoprothiolane	Rice blast	Japan
	Validamycin A	Rice sheath blight	Japan
1980s	Coratop	Rice blast	Switzerland and Japan

Thus, we have to direct our attention to developing agrochemicals which have no risk to all mankind and the environment. We would like to consider how to achieve this purpose.

Before we enter into the discussion, let us look at the history of the development of fungicides.

As shown in Table 1, fungicides were used for seed disinfection and prevention of wood decay in the 18th century. In the 19th century, the lime-sulfur and Bordeaux mixtures were developed as spray fungicides for rust and powdery mildew disease of cereals, and downy mildew of the grape vine. Organomercurials were used as a seed disinfectant in the beginning of the 20th century; and these compounds were found to be effective in controlling the blast disease of rice plants in the 1950s in Japan and were widely used all over the country. In those days scientists focused attention on killing the pathogenic fungus but did not pay attention to the risks. This might be due to the fortuity that the lime-sulfur and Bordeaux mixtures did not cause problems. However, later on organomercurial residuals were found in rice and accumulated in the bodies of those who ate the sprayed rice. The use of mercuric compounds as pesticides is prohibited by the Japanese government now. These circumstances made Japanese scientists pay attention to the toxicity of fungicides against human health. As a result, selective control agents for the rice blast disease such as blasticidin S, kasugamycin, and organophosphorus compounds have been developed. Benomyl, developed in the United States, is a potent fungicide which shows selective toxicity to many plant pathogens with low toxicity to higher plants and animals.

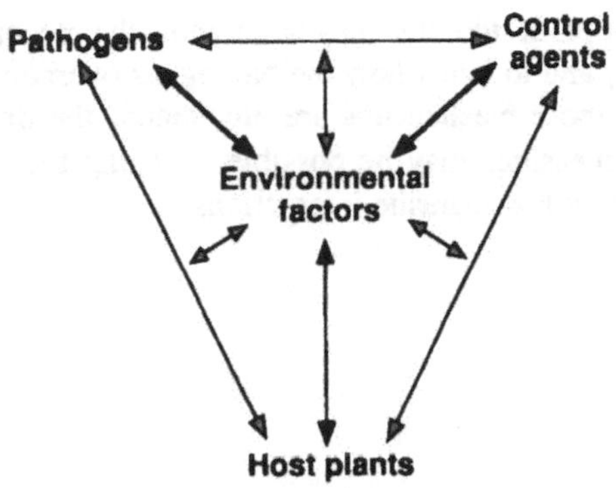

FIGURE 1. Interrelationships of plant disease control by pesticides.

In order to avoid injurious effects to higher plants and animals, scientists examined compounds having selective toxicity. This caused the appearance of tolerant strains of pathogenic fungi against fungicides.

Since the 1970s in Japan, many nonfungicidal disease control agents have been developed against the rice blast disease. This turning point was made by research conducted by myself[13] (1968) on the mechanism of the blast control activity of pentachlorobenzyl alcohol. This compound does not inhibit germination of spores or mycelial growth of the blast fungus even at a concentration of 1000 ppm, but does inhibit penetration into the host plant at a concentration of 50 ppm.

By this study, it was demonstrated that compounds which have no antifungal activity can be useful as disease control agents; and scientists involved in pesticide development came to recognize the importance of studying the host-parasite interaction, especially the mechanism of "pathogenicity".

In order to control plant disease by agrochemicals, scientists have to keep in mind that environmental conditions deeply affect the basic factors, that is, higher plants, pathogens, and chemicals. The situation is not as simple as killing the pathogen with a fungicide, but is much more complicated (as shown in Figure 1).

From the standpoint of plant pathology, chemical compounds for the control of plant diseases should have one of the following three properties:

1. Exhibit fungicidal or fungistatic activity
2. Be able to inactivate the pathogenicity
3. Be able to activate the resistance response of the host plant to pathogens

The above three principles are effective not only in chemical control but also in biological, physical, and integrated control of plant diseases.

The starting point for plant disease control is to know the mechanism of disease outbreak, in other words, the understanding of what the pathogenicity is. From the host plant side, this may be to know the mechanism of resistance against diseases, and to know how the pathogens overcome the resistance of hosts. After all those mechanisms are elucidated, the development of new disease control measures may be possible by using the natural phenomena which occur within host-parasite interactions.

REFERENCES

1. Stakman, E. C., The role of plant pathology in the scientific and social development of the world, *Plant Pathology, Problems and Progress*, Holton, C. S., Fisher, G. W., and MacCallan, S. E. A., Eds., The University of Wisconsin Press, Madison, 1959, 3.
2. Hino, I., *Developmental History in Plant Pathology*, Asakura, Tokyo, 1949, 115.
3. Department of Statistics, Ministry of Agriculture, Forestry and Fishery of Japan, *Statistic Analysis of Agricultural and Forest Products*, Tokyo, 1949–1962.
4. Takai, S., Elm-killer — Dutch elm disease, *Kagaku To Seibutsu*, 22, 462, 1984.

4a. Smalley, E. B., personal communication.

5. Mamiya, Y. and Kiyohara, T., Description of *Bursaphelenchus lignicolus* n. sp. (Nematoda: Aphelencoidae) from pine wood and histopathology of nematode infected trees, *Nematologica*, 18, 120, 1972.
6. Nickle, W. R., Golden, A. M., Mamiya, Y., and Wergin, W. P., On the taxonomy and morphology of the pine wood nematode, *Bursaphelenchus xylophilus* (Steiner & Buhrer 1934) Nickel, 1970, *J. Nematol.*, 13, 385, 1981.
7. Steiner, G. and Buhrer, M., *Aphelenchoides xylophilus* n. sp., a nematode associated with blue-stain and other fungi in timber, *J. Agric. Res.*, 48, 943, 1934.
8. Mamiya, Y., Finding of pinewood nematode in the U.S., *For. Pest*, 29, 54, 1983.
9. Dropkin, V. H., Linit, M., Kondo, E., and Smith, M., Pine wilt associated with *Bursaphelenchus xylophilus* (Steiner & Buhrer, 1934) Nickel, 1970, in the United States of America, in *Proc. XVII IUFRO World Congr. Div.*, 2, 1981, 265.
10. Theophrastos of Eresos (cf. Kirchiner, O., Die botanischen Schriften des Theophrast von Eresos, *Jahrb. Klass Philologie*, 7, Suppl. 449, 1978).
11. Flor, H. H., The complementary genetic systems in flax and flax rust, *Adv. Genet.*, 8, 29, 1956.
12. Marasas, W. F. D. and Nelson, P. E., *Mycotoxicology*, Pennsylvania State University Press, University Park, 1987, 28.
13. Oku, H. and Sumi, H., Mode of action of pentachlorobenzyl alcohol, a rice blast control agent — Inhibition of hyphal penetration of *Pyricularia oryzae* through artificial membrane, *Ann. Phytopathol. Soc. Jpn.*, 34, 250, 1968.

CHAPTER 1

Pathogens and Pathogenicity

I. EVOLUTIONAL ASPECTS OF PLANT PATHOGENS

According to Agrios,[1] about 100,000 species of fungi and 1600 species of bacteria are living on this planet; and among these, about 8000 species of fungi and 200 species of bacteria attack and injure higher plants. Another 100 species of these microorganisms cause diseases in animals. Most of the 100,000, and 1600 species of fungi and bacteria are strictly saprophytic, living on nonliving organic matter and decomposing enormous quantities of animal and plant remains. In other words, saprophytic microorganisms are scavengers of our environment, recycling dead organic matter into simple elements.

Thus, the nutritional procedure of pathogenic microorganisms that derive food materials from plants and animals is the exceptional case in the kingdom of microorganisms. Because most microorganisms are saprophytes, they live in violent competition with each other, seeking food materials and living space. The production of antibiotics is likely to be one of the strategies of microorganisms to defend their living territory.

Ogura[2] examined the change of microflora on rice straw added in the soil and found that the fungi with fermentation ability came first, and then the fungi were altered by cellulose-decomposing fungi. This alteration seemed to be due to both the production of antibiotics by the latter fungi and the exhaustion of nutrients for the former fungi. These cellulose-decomposing fungi persist for a long time; but when the cellulose is exhausted, these fungi disappear and then lignin-decomposing fungi appear and multiply on it. This model experiment shows the violent competition between soil microorganisms, and the complete decomposition of organic substances by a close cooperation of many microorganisms.

Among these saprophytic microorganisms which compete with each other if one acquires the ability to derive food materials from a plant and multiply on it, the microorganism can escape from the violent competition. The plant pathogenic microorganisms are likely to be developed in such a way from saprophytic ones by parasitic adaptation. This hypothesis might be supported by the evidence that there are varieties of pathogenic fungi of which parasitism is in varying degrees, between obligate parasitic and facultative parasitic. The facultative parasite, which usually lives as a saprophyte but under some conditions can parasitize on a plant, seems to be a pathogen with the lowest parasitic adaptation. On the contrary, the uppermost parasitic adaptation can be seen in the obligate parasite which can derive food materials only from living plant cells but not from dead cells.

Nelson,[3] however, hypothesized that fungi had originated from obligate parasitic microorganisms and evolved in the direction of saprophytes. He considered that primitive fungi could not live separately from living plants; however, after acquiring the abilities to produce dormant spores, they became able to live independently from plants as a saprophyte. However, regarding the parasitism of plant pathogenic fungi, he considered that saprophytes regained parasitic ability acquiring beneficial properties from both parasitic and saprophytic life. Obligate parasites, in this sense, completely regained the ancient lifestyle.

The parasitic adaptation for microorganisms to be plant pathogens includes acquisition of the following three abilities:[4] (1) ability to enter into a plant, (2) ability to overcome the resistance of the host, and (3) ability to evoke disease.

These abilities are essential to plant pathogens, and are not found in saprophytic microorganisms. From these standpoints, many steps of adaptation are found in those three properties in each plant pathogen. In fact, there are varieties of plant pathogens of which parasitic adaptations are different. Relationships between the degree of parasitic adaptation and the three properties of pathogens are indicated in Table 1.

As described, the pathogens of which parasitic adaptations are low usually live as saprophytes on plant and animal remains, but under certain conditions invade plants. These pathogens are called facultative parasites. Many facultative parasites enter into plants through wounds, and have no special way of entering. These pathogens mainly parasitize on organs or tissues in which physiological activity is low, such as fruits and other storage organs, trunks of trees, or vascular bundles. Several fungi which cause damping off of young seedlings belong to facultative parasites and generally have wide host ranges.

On the contrary, obligate parasites such as powdery mildew, downy mildew, rust fungi, and so on seem to have lost the saprophytic ability. For these fungi, death of the host cell results in death of the pathogen itself.

Table 1. Classification of Plant Pathogenic Fungi by Degree of Parasitic Adaptation and Their Properties

Classification	Parasitic adaptation	Mode of infection	Colonized on	Factors to suppress host defense	Factors to evoke disease
Saprophytes	None	—	—	—	—
Facultative parasites	Low ↑	Wound	Physiologically inert organs	NST, enzymes	NST, enzymes
Perthophytes	↕	Proper ways Stomato, Cuticula, Gland etc.	Physiologically inert and active organs	NST, HST, enzymes, suppressors	NST, HST, enzymes
Facultative saprophytes	↕	(as above)	(as above)	Suppressors	Complicated, weak or no factors from pathogens
Obligate parasites	↓ High	(as above)	(as above)	Suppressors including endogenous ones	(as above)

Note: NST: nonspecific toxin, HST: host specific toxin.

Obligate parasites have their own procedures for entering into the host plant. The stomatal penetration by rust fungi and downy mildew fungi is likely to be one of the most efficient entrance methods. The tip of the germ tube from the germinated uredospore comes over a stoma, enlarges, and becomes closely appressed to the surface of the guard cells. From the structure thus formed an appressorium, a very fine penetration hypha emerges and enters through the stomatal pore into the substomatal cavity. In downy mildew fungi, the motile zoospore swims in the water film on the leaf surface. When it reaches a stoma, its movement appears subject to some stimulus and the spore comes to rest over the stomatal pore. From the now nonmotile spore, a slender hypha emerges and passes through the pore into the substomatal cavity. The closed stoma can be forced to penetrate through this type of penetration. For fungi which do not form an appressorium and enter through the stomata directly by hyphae like *Cladosporium fulvum*, the closed stomata seems to contribute a barrier through which this type of growth cannot pass. These facts suggest that the type of stomatal entry shown by rust uredospores and downy mildew cystospores is more likely to be effective.[5]

The other invading procedure for obligate parasite is so-called direct penetration, by which an unbroken plant surface is penetrated. These fungi do not make use of either wound or natural openings.

Obligate parasites seem to have some mechanisms to overcome the defense reaction of the host without killing host cells. Other toxic metabolite or deleterious enzymes for the host cell should not be produced by obligate parasites because this type of pathogen cannot live on dead cells as described previously.

The virulence factors in obligate parasitic diseases may not be due to the direct action of the pathogen, and the mechanism of symptom development seems to be more complex than the other types of diseases. Because obligate parasites cannot live saprophytically, they should overcome adverse conditions by producing dormant organs or by altering the host plant.

Between these two extremes, perthophyte and facultative saprophytes exist. The perthophyte is the pathogen which kills the host cell by toxins or enzymes, and then lives saprophytically on the dead cell. The facultative saprophyte is the pathogen that parasitizes usually on a living host cell, but can live saprophytically under certain conditions.

The pathogen of *Helminthosporium* leaf blight disease, *Cochliobolus miyabeanus;* pathogens of many black spot diseases caused by *Alternaria* spp.; and pathogens of wheat stripe disease, *Cephalosporium gramineum*, belong to the perthophytes. Many perthophytes produce necrotic symptoms. The perthophytic fungi also have their own way of entering into plants, direct penetration or penetration through natural openings.

The perthophyte produces toxins or deleterious enzymes to overcome the resistance of host plants. These fungi can parasitize on the physiologically active parts of host plants. The host range of the perthophyte is divergent, that is, narrow for some fungi but wider for other fungi. The fungi which have a wide host range generally produce nonspecific toxins or deleterious enzymes to plants; however, some fungi which produce host-specific toxins have a narrow host range, and some parasitize on one species of plants or even on only susceptible cultivars of the species. Further, some perthophytes produce substances that suppress the elicitation of defense reactions of the hosts, which do not give the visible injury; and after establishing infection, they produce another substance to cause necrotic symptoms. For these fungi, the host range is generally narrow.

The facultative saprophyte is situated between the perthophyte and the obligate parasite. Pathogens of many smut diseases and the leaf curl disease of the peach caused by *Taphrina deformans* seem to belong to this group. The facultative saprophyte also has its own way of entering into host plants, direct penetration or stomatal penetration; and scarcely produces deleterious toxins or enzymes. Many of these types of parasites seem to have some mechanisms to suppress the resistance reaction of their hosts.

Here I have classified plant pathogens into four types according to the parasitism; however, in nature the parasitic fungi cannot be divided neatly into four types as I described.

In the higher plant kingdom, prosperity and decay of species might occur during the coevolution with plant pathogens. Some species might be exterminated by the attack of vigorous pathogens, others could overcome these pathogens and preserve their species, and yet others might obtain genes for resistance by hybridization between related species. In other words, at the present time, both pathogens and host plants have built up an atmosphere of coexistence under the natural ecosystem. This view may be supported by the fact that new pathogens introduced from foreign countries cause serious problems in many areas. The plant quarantine system in many countries is aimed at inhibiting such a tragic problem in native plants which do not have resistance genes against new pathogens.

II. VARIATION AND SPECIALIZATION OF PATHOGENICITY

Breeding a resistant cultivar or variety is one of the effective control measures for plant diseases. However, occasionally the newly developed resistant variety against some pathogens suddenly becomes susceptible. The cause of this phenomenon is due to the variability of pathogenicity of the pathogens.

When some groups of individuals within one species of a pathogen show pathogenicity to a different genus or species of plants, the groups are called *formae speciales*. When individuals within the same *forma specialis* obtained the pathogenicity to the restricted varieties or cultivars of a species of host plant, the group of such individuals are called physiological races or races. The phenomenon of the appearance of new races is called physiological specialization.

Many races are found within the pathogens of improved crop plants which are important for mankind. This means that variation or specialization of the pathogenicity is a man-made phenomenon, namely, induced by breeding of new varieties or cultivars.

In general, when a new resistant cultivar or variety is developed, the new race of pathogen which acquires pathogenicity to this cultivar or variety appears within several years. The old races avirulent to the newly developed cultivar become extinct because they cannot parasitize, and thus the alternation of races occurs.

The appearance of new races described above is a typical example of parasitic adaptation which occurs during a very short period of time. Therefore, it is possible that the variation of pathogenicity occurs against the genus or family level of higher plants, and further that saprophytes might have acquired pathogenicity to higher plants during the long period of evolutionary process.

III. FACTORS CAUSING VARIATION OF PATHOGENICITY

The mechanisms on physiological specialization (variation of pathogenicity) will be considered here. Any one of a number of mechanisms that can result in genetic recombination may lead to the variation of pathogenicity. The hybridization of different races or formae speciales may be one of the causes for the variation of pathogenicity. For example, by crossing different races of flax fungus, ***Melanpsora lini***, many races which have different pathogenicity are segregated at the F2 stage.[6] The powdery mildew fungus of cereals, ***Erysiphe graminis***, has many specialized forms which parasitize on different species of gramineous plants such as wheat, barley, and agropyron; and these specialized forms can be hybridized by each other.[7] Some of the progenies of the hybridization between wheat fungus and agropyron fungus can parasitize both on wheat and agropyron, but the pathogenicity is lower than the parent fungi.[8]

The variation of pathogenicity of some fungi which have no perfect stage (sexual stage) may occur through mutation, heterocaryosis, and parasexualism.

Mycelia and spores of fungi sometimes have more than two nuclei in one cell. Many of such fungi have the same genetic type of nuclei in each mycelial cell. These mycelia are called homocaryon. In contrast, mycelia which contain different types of nuclei are called heterocaryon; and the phenomenon is called heterocaryosis. Heterocaryosis occurs by fusion between hyphae which contain different types of nuclei, and as a result the exchange of genes for pathogenicity may occur. The variation of pathogenicity of the rice blast fungus has been believed to occur by this way. The other possibility of the appearance of a new race may be through parasexualism. The nuclei of imperfect fungi are usually haploid, but sometimes become diploid in mycelia by the fusion of two haploid nuclei.

The susceptibility or resistance of host and virulence or avirulence of the pathogen are determined by gene-for-gene interactions between both organisms;[9] and the mutation or recombination of genes for virulence of pathogens is responsible for the variation of pathogenicity of pathogens to the cultivar, variety, or even species of the host plants. At the host-parasite interface, the interaction of some substances, gene products, from the host and parasite actually determines the susceptibility or resistance of hosts and pathogenicity of the pathogens.

Though appearance of new formae speciales and races are due to the recombination or mutation of genes for virulence of the pathogen, many properties other than pathogenicity of the pathogen may vary as a result of those nuclear behaviors. Among variants, the one in which direction of the variation does not adapt to survival, reproduction, or environmental condition may become extinct.

The fate of new races and formae speciales of the pathogen of cultivated plants is affected largely by the variety or cultivars of plants grown in each area. In areas where only tolerant plants or nonhost plants are cultivated, the new race or forma specialis disappears because of the lack of host plants (nutrients). On the contrary, if we introduce a new resistant variety into a field with a small population of variant pathogens that can invade the new variety, the variant multiplies without any competition and produces a large population; and thus the alteration of race occurs. As a result, the variant destroys the new resistant variety. Therefore, we must examine carefully the distribution and existence of races in the area before the introduction of a new resistant variety.

IV. WHAT IS PATHOGENICITY?

A. Pathogens and Pathogenicity

There are many definitions of plant disease. For example, Westcott[10] agreed with the definition by Whetzes, that disease in a plant is an injurious physiological process caused by the continued irritation of a primary factor, exhibited through abnormal cellular activity and expressed in characteristic pathological conditions called symptoms. Horsfall and Dimond[11] defined disease as a malfunctioning process that is caused by continuous irritation; hence disease is a pathological process, and this conception was accepted by the Committee of Terminology of the American Phytopathological Society and by its counterpart committee of the British Mycological Society. However, Agrios'[12] definition is more acceptable. That is, he described that a plant is healthy or normal when it can carry out its physiological functions to be the best of genetic potential. These functions include normal cell division, differentiation and development, absorption of water and minerals from the soil and translocation of these throughout the plant, photosynthesis and translocation of the photosynthetic products to areas of utilization or storage, metabolism of synthesized compounds, and reproduction and storage of food supplies for overwintering or reproduction. Briefly speaking, when the nutrition, growth, and reproduction — the characteristic genetic functions of the individual species — are kept normal, the plant is healthy. Therefore, whenever plants are disturbed by pathogens or by certain environmental conditions and one or more of these functions are interfered with beyond a certain deviation from the normal, then the plants become diseased.

Therefore, automatically the term pathogen can be defined as "internal and external factors which cause abnormality on the physiology of the above three processes". The following are considered to be pathogens:

1. Nonliving pathogens consist of soil conditions such as humidity, temperature, pH, physical structure, lack of oxygen, fertilizer, toxic substance, etc., and climatic conditions such as light, temperature, humidity, wind, rain, snow, thunder, etc. Injury by agriculture practices such as machines, phytotoxicity of agrochemicals, industrial by-products such as air and water pollution, etc., and plant metabolites are also factors.
2. Living pathogens are animals such as insects, nematodes, mites, higher animals, etc., and plants such as mycoplasma, slime mold, bacteria, fungi, and parasitic higher plants.
3. Viruses and viroids are also pathogens.

Among these pathogens, diseases caused by nonliving pathogens are not contagious and are called nonparasitic, noninfectious, or physiological diseases. In contrast, diseases caused by living pathogens, viruses, and viroids are contagious and are called parasitic, or infectious diseases. Pathogens which are dealt with in the field of plant pathology are, generally, mycoplasma, bacteria, fungi, and nematodes among the living pathogens in addition to viruses and viroids. Of all the infectious pathogens, fungi are known to cause the most serious injury on plants.

The infectious pathogens have the ability to invade plants and interfere with their function, and this ability is called pathogenicity. For pathogenicity, the following three abilities are essential. One is the ability to enter into plants, the second is the ability to overcome the resistance of host plants, and the third is the ability to evoke disease as described earlier.

B. Entry Into Plants

Plant pathogens have to enter into plants in order to derive the food materials they require. They do this by one or more of the following four ways: entry through wounds, entry through natural openings, direct penetration, and penetration through localized organs.

1. Entry Through Wounds

In general, pathogenic fungi of forest trees and those causing rot of fruits and other storage organs enter into plants through wounds. Plant pathogenic fungi, of which parasitic adaptation is not as advanced, commonly use this mode of entry. For example, tissues of the trunk of trees except cambium, phloem, etc. are composed of dead cells; and the storage organs such as fruits and tubers are physiologically not as active compared with the other growing organs such as leaves and stems. Therefore, primitive pathogenic fungi can only enter into plants through wounds and live in these organs. Because these fungi cause serious crop losses, there is a great deal of literature on what the types of wounds are and how these wounds can be avoided. Many fungi (species of *Polyporus, Polia, Fomes,* and *Ustilina*) which cause heart rot of standing timbers enter through wounds made by natural phenomena such as fire or other weather conditions.[13] The wounds that arise as a result of agricultural practices such as pruning[14] and grading of products[15] are also the entry sites of these pathogens.

In addition the wounds made by insects are entry sites for some pathogens. In the Dutch elm disease caused by *Ceratocystis ulmi,* spores of this fungus are transmitted by the bark-boring beetle, *Scolytidae;*[16] when the beetle bores into the trunk of an elm tree, some spores attach on the cavities made by the beetle some reaching to the xylem vessel, are transmitted throughout the whole tree by the transpiration stream. Further, the bark beetles lay eggs in the cavities, hatch, and grow into pupae and adults under the bark. When these adult beetles escape through perforated holes which are made by their mother beetle, the spores of pathogenic fungus produced on the surface of holes attach to the young adult and are transmitted to other elm trees.

The wilting disease of the Japanese pine tree caused by the pinewood nematode, *Bursaphelenchus xylophilus,* is transmitted in a way similar to the pathogen of Dutch elm disease. The pine sawyer, the genus *Monochamus,* grown in the pine trunk emerges from the pine tree with many pinewood nematodes in and out of the body. The newly emerged adults attain maturity mainly by feeding off the soft tissues of living branches of pine trees, when the pathogenic nematodes attached to the pine sawyer enter into the new shoots through wounds.[17]

Because viruses are nucleoproteins and nonliving agents, they have to be transmitted by other living organisms, fungi, nematodes, and especially by insects such as aphids and hoppers, entering into host plants through wounds made by such organisms.

Bacterial pathogens such as *Agrobacterium tumefaciens* — the crown gall pathogen — and *Erwinia carotovora* which causes soft rot of vegetables also enter into plants through wounds. Because bacteria are lower organisms as compared with fungi, the parasitic adaptation is likely to be lower than the pathogenic fungi. For fungi and bacteria which enter through wounds, the old wounds are a less favorable entry site because healing compounds such as cork, gum, and tyrosis are formed and inhibit the penetration.[18]

2. Entry Through Natural Openings

There are many openings such as stomata, hydathode, and lenticels on the surface of plants. Among bacterial pathogens, the fire blight pathogen of the

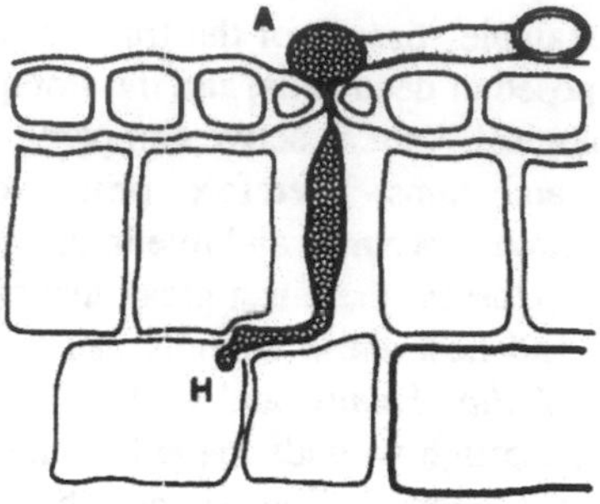

FIGURE 1. Schematic illustration of stomatal entry. (A) Appressorium; (B) haustorium.

apple, *Erwinia amylovora*, enters through lenticels; and the bacterial blight pathogen of rice, *Xanthomonas campestris* pv. *oryzae*, enters through the hydathode.

The process of entry through stomata by pathogenic fungi involves two phases (Figure 1). The first phase is the elongation of the germ tube to stomata, and the second phase is the growth through the stomatal pore. It seems that no direct stimulus may be involved[19] for the first phase because germ tubes of many fungi on the host surfaces grow in a random direction.[20] On the other hand, for the second phase it can hardly be considered accidental, particularly in those fungi which form appressoria and penetration hyphae. The hyphae changing course to a vertical direction is considered to be due to the stimulus of water vapor raised through stomata. That is, Balls[21] perforated the thin membrane of Indian rubber comparable to the size of stomata. The membranes were arranged with one side exposed to air saturated with water vapor and the other side to the laboratory air. The latter side was seeded with uredospores of *Puccinia glumarum*. Two days later, many germ tubes were found entering through the pore, but none grew back to the side exposed to laboratory air. In this experiment, he could not find appressorium in the perforated hole. In the host cell, uredospores of many rust fungi and zoospores of many downy mildew fungi form appressorium. For those fungi, there must be some stimuli to form appressoria. The contact stimulus is considered one of the stimuli forming appressoria. For example, Dickinson[22] found the formation of appressorium-like structures when uredospores of several rust fungi are germinated on the paraffin-wax-collodion membrane which is incorporated with cell wall fragments. The effect of chemical stimuli seems to be also involved in the formation of appressoria because uredospores of many rust fungi produced appressoria-like structures when germinated in nutrient agar but not in water agar.[23]

3. Direct Penetration

Many plant pathogenic fungi enter into plants by penetration through an unbroken plant surface, in particular the cuticle which covers the surface.

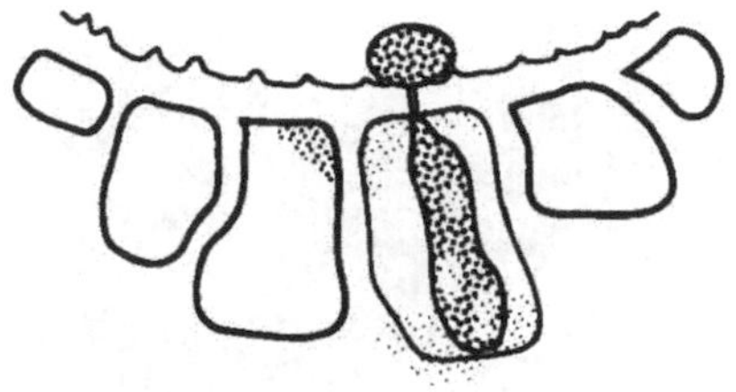

FIGURE 2. Direct penetration of motor cell of rice leaf by *Helminthosporium* leaf blight fungus.

This type of entry is called direct penetration or cuticular penetration. Generally, pathogenic fungi entering by direct penetration do not make use of either wounds or natural openings as entry sites.

The way of direct penetration has been studied for many host-parasite combinations, e.g., *Phytophthora infestans* and potato,[24] *Venturia inaqualis* and *V. pirina* on apple and pear,[25] *Erysiphe graminis* on wheat,[26] *Pyricularia oryzae* on rice,[27] *Cochliobolus miyabeanus* on rice,[28] and so on (Figure 2). The essential features in the infection process for these host-parasite combinations follow a similar pattern. First, the germ tube arising from conidia becomes flattened or forms appressorium when contacting the cuticle surface. Second, from structures thus formed fine penetration hyphae emerge and penetrate through the epidermal wall. Penetrated hyphae then increase in diameter of normal size or form haustoria.

Mechanisms of direct penetration were reviewed by Wood[29] and Dickinson,[30] and the importance of chemical and physical factors has been the subject of discussion. Many pathogenic fungi which invade plants by direct penetration have been reported to penetrate not only the epidermal wall of nonhost plants but also artificial membranes. The germ tube of rice blast fungus, *Pyricularia oryzae; Helminthosporium* leaf spot fungus of rice, *Cochliobolus miyabeanus;* and black rot fungus of citrus, *Alternaria citri* can penetrate through a cellophane membrane.[31] The germ tube of pathogenic fungi belonging to the genus *Botrytis* can penetrate through a membrane of polyvinylformal.[32] By adjusting the thickness of the membrane, one can determine the penetrating ability of pathogenic fungi. These facts suggest that the physical factor (mechanical puncture) is predominant in the direct penetration, because these artificial membranes are unlikely to be affected enzymically. Further, the ability of certain fungi to infect is directly related to resistance of the outer tissue to mechanical puncture, e.g., *Puccinia graminis* on barberry[33] and *Pyricularia oryzae* on rice leaves.[27] In spite of this evidence that the physical factor is predominant in direct penetration, certain cytological evidence suggests that enzymatic degradation of the host surface may play a part in penetration. Figure 3 shows the electron micrograph of the penetration site of powdery mildew fungus, *Erysiphe graminis,* into a barley leaf. One can notice that the epidermal layer of the host plant contacting the appressorium seems to be somewhat degraded in advance to penetration by the fungus.[34]

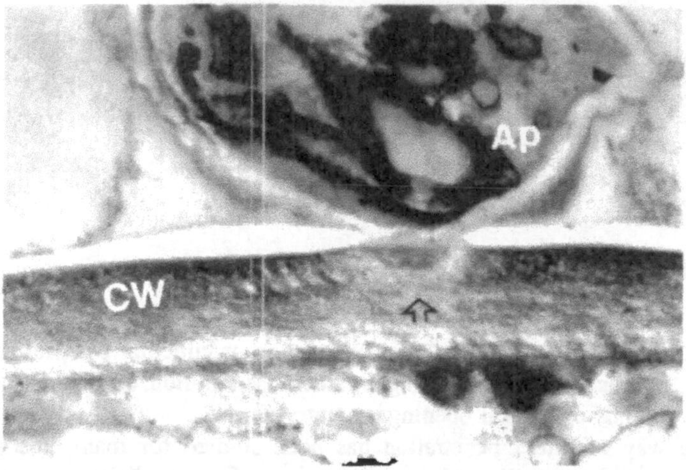

FIGURE 3. Early stage of infection by powdery mildew fungus on barley leaf. Ap: Appressorium of powdery mildew fungus, CW: host cell wall, Pa: papilla.

The role of polymer degrading enzymes, especially cutinase in direct penetration, was suggested and reviewed by Kolattukudy[35] and Kolattukudy and Crawford.[36] They demonstrated by a scanning electron microscope that cutinase was present where penetration of *Fusarium solani* f. sp. *pisi* was occurring to a pea epidermis with ferritin conjugated anticutinase immunoglobulin. They also found that the inhibition of cutinase by antibody raised against cutinase or specific chemical inhibitors prevented fungal penetration into the plant. They further suggested the importance of suberin- and pectin-degrading enzymes in fungal pathogenicity.[36]

These conflicting experimental results suggest that the importance of physical or chemical factors may differ in each host-parasite combination. Further, in many cases, both physical and chemical factors act cooperatively for direct penetration at the actual site of host-parasite interface.

Because penetration of the rice blast fungus through a cellophane membrane is activated by adding the rice straw extract beneath the cellophane,[31] some components of a plant may be acting as the stimuli for direct penetration. This assumption is also supported by evidence that the conductivity of water droplets increases when placed on the leaf surface of many plant species, and the germination of spores of plant pathogens is activated in these droplets.[37] *Rhizoctonia* sp. and *Helicobasidium* sp., which cause damping off of young seedlings, usually form an infection cushion composed of layers of mycelia before the hyphal penetration. The formation of an infection cushion is also stimulated by some components of the host diffused from underlying cells.[38] Usually young seedlings are the subject of direct penetration.

The rice blast fungus[39] and *Helminthosporium* leaf blight pathogen[27] can invade adult rice leaves, but the site of penetration is mostly the motor cell (see Figure 2). This seems to be due to the low deposition of silica and thinness of the epidermis of the motor cell. Thus, the weakly developed epidermis is the favorable entry site for direct penetration.

4. Penetration Through Localized Organs

Plant surfaces have several organs in which protective tissues are not well developed, e.g., stigma, glands, hair roots, and so on. Infection of cereal plants by several smut fungi such as *Ustilago nuda* and *U. tritici* occurs at anthesis. Spores of these fungi germinated on the stigma and the germ tubes pass into growing points of the embryo.[40] Therefore, seeds developed from such embryos have the pathogen inside, but still have the capacity to germinate.

Sclerotinia mali causes leaf, flower, fruit, and stem rot on the apple plant. Of these, fruit rot is caused by infection of the ovule through the stigma at anthesis.[41]

In some flower infections, insects play a very important role as vectors. For example, the fire blight bacterium, *Erwinia amylovora,* is transmitted by insects and windblown rain.[42] The conidia of ergot fungus, *Claviceps purpurea,* are also transmitted on floral organs by insects from blossoming cereals and grasses to maturing crops.[43] Thus, plant pathogens might acquire the ability to enter into plants by ways described above, and these properties might be the first requisite to becoming plant pathogens.

C. Suppression of Defense Reaction of Host and the Host Range

After pathogenic propagules enter into plants by penetrating through the defense barrier of the surface, their next important task is how to overcome the active defense mechanisms expressed by host plants.

Yoshii[44] reported that the rice blast fungus inoculated onto the tomato leaf was killed immediately after penetration (Figure 4). This fact clearly indicates that the blast fungus is a pathogen of rice and has an ability of direct penetration into plants, but does not have the ability to overcome the defense reaction of the tomato plant.

Most plant pathogens have their own host plants, and the plant species to which each pathogen can parasitize is called the host range. Some plant pathogens can parasitize on a wide range of plants, but some can infect only one species or even a variety of a cultivated plant species. Thus, the phenomenon that a pathogen has its own host plants is called host specificity. The host specificity and host range of each pathogen are determined by plant species of which defense reactions are overcome by the pathogen. The host specificity of the pathogen is likely to be determined during the process of coevolution with plant species.

As to the actual strategies of a pathogen for overcoming a host-defense mechanism, nonspecific toxins and deleterious enzymes secreted by pathogens may be involved in primitive pathogens. That is, they overcome host defense by killing the host cell and live on them saprophytically. However, in

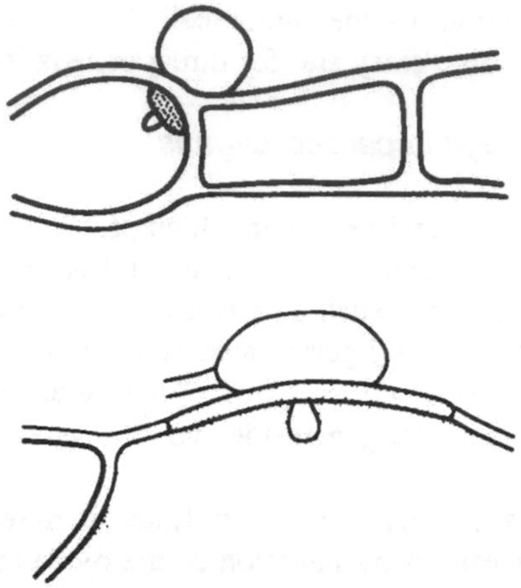

FIGURE 4. The rice blast fungus inoculated on tomato leaf is killed immediately after the penetration through epidermis. (From Yoshii, H., *Ann. Phytopathol. Soc. Jpn.*, 18, 14, 1948. With permission.)

pathogens as highly developed as obligate parasites, they must have some other mechanisms to suppress the expression of defense reaction of host cells without killing, because the death of the host cell results in the death of the parasite itself. This mechanism will be described in Chapter 3 in detail. The explanation of this problem is the main purpose of this book.

D. Factors to Evoke Disease (Virulence Factors)

Microorganisms which enter into plants and establish only the nutritional relationship are not pathogens. For example, root nodule bacteria parasitize on roots of leguminous plants, fix the nitrogen in the air, and supply it to host plants; and in return they derive nutrients from hosts to survive and multiply. This relationship is not disease but symbiosis. In some insects such as aphids, the symbiotic microorganism is essential to support their life;[45] therefore, the elimination of the symbiont results in the death of the insect. The research of the bactericide against such symbiotic bacterium is now one of the targets for insect control. That is, the bactericide becomes an insecticide.

Main problems caused by pathogens in cultivated plants are the decrease in yield and the deterioration of quality. These deleterious effects may be caused by the following factors: deleterious enzymes, toxin produced by pathogens, and abnormal metabolites resulting from host-parasite interactions.

1. Enzymes as Virulence Factors

Many plant pathogens grow within the host plants intercellularly by breaking down the middle lamellae with pectolytic enzymes. These enzymes sometimes play roles in symptom development. Anthracnose of the persimmon has two phases in pathogenesis, dry rot and soft rot. According to Tani,[46,47] the symptom dry rot appearing in the early stage of pathogenesis is caused by the pectolytic enzymes produced by the *Gloeosporium kaki* pathogen; and the later stage symptom, the soft rot, is caused by the tissue-macerating enzymes originating from host fruits produced by the stimuli of infection. Soft rot of vegetables caused by bacterial pathogens is believed to be a result of pectolytic and cellulolytic enzymes bacterial in origin.[48,49] As to the pectolytic enzymes in bacterial pathogens, depolymerase seems to be involved in soft rot because *Erwinia carotovora, E. atroseptica,* and *E. aroideae* do not produce pectin methyl esterase and polygalacturonase but produce abundant depolymerase in the medium supplemented with pectin.[50]

According to Basham and Bateman,[51] endopectate lyase (endopectate transeliminase) might be the main pectic enzyme for tissue maceration because the purified enzyme caused the maceration and cell death of potato and tobacco pith tissue. *Erwinia chrysanthemi* was reported to produce isozymes of endopectate lyase with acidic, neutral, and alkaline isoelectric points in culture. Each of these isozymes macerates plant tissue, but the alkaline isozyme has the highest activity; hence it is the most destructive to plant tissues.[52]

Notwithstanding these reports, the artificial mutant of *E. chrysanthemi* (which was produced by genetic engineering) deleting all genes encoding pectate lyase isozymes, still has macerating activity in potato, carrot, and pepper tissues.[53] This suggests the possibility that factor(s) other than pectate lyase may be involved with the tissue maceration.[54]

Two kinds of wood decay caused by wood rotting Basidiomycetes, white rot and brown rot, and the mechanism to cause these rots are summarized by Agrios.[1] The former is caused by white rot fungi. They enzymatically decompose lignin and hemicellulose first and cellulose last, or decompose all wood components simultaneously. As a result, a light-colored spongy mass (white rot) remains. On the contrary, brown rot is caused by fungi which preferably attack soft wood; they degrade and utilize primarily the cell wall polysaccharide, cellulose, and hemicellulose, leaving brown lignin. For the lignin-degrading ability, polyphenoloxidases of the laccase type produced by white rot fungi play important roles.[55]

One of the symptoms of the *Helminthosporium* leaf blight disease of rice is formation of small necrotic spots on leaves. These necrotic spots are formed by the very active, stable polyphenoloxidase activity of the pathogenic fungus.[56] In more detail, polyphenols remarkably increased by the infection with the fungus, especially in the cells adjacent to the infected cells. The increased polyphenols are then oxidized, and the oxidation products are polymerized and/or condensed with amino acids into brown substances.[28] The oxidation products of polyphenols inhibit the growth of the pathogenic fungus; hence brown spots remain small.[57,58] Thus, the enzymes produced by pathogenic fungi play important roles as factors to evoke diseases in many host-parasite combinations.

Table 2. Plant Pathogenic Fungi Which Produce Host-Specific Toxins

Fungi	Host	Toxin	Site of action
Alternaria alternata apple pathotype	Apple	AM	Plasma membrane Chloroplast
A. alternata Japanese pear pathotype	Japanese pear	AK	Plasma membrane
A. alternata rough lemon pathotype	Rough lemon	ACR	Mitochondria
A. alternata strawberry pathotype	Strawberry	AF	Plasma membrane
A. alternata tangerine pathotype	Tangerine	ACT	Plasma membrane
A. alternata tobacco pathotype	Tobacco	AT	Mitochondria
A. alternata tomato pathotype	Tomato	AL	Mitochondria
Corynespora cassiicola	Tomato	CC	?
Helminthosporium carbonum	Corn	HC	Plasma membrane
H. maydis race T	Corn	HMT	Mitochondria
H. sacchari	Sugarcane	HS	Plasma membrane
H. victoriae	Oat	HV	Plasma membrane
Perconia circinata	Sorghum	PC	Plasma membrane
Phyllosticta maydis	Corn	PM	Mitochondria

2. Toxins as Factors to Evoke Diseases

Damages and symptoms of many plant diseases are caused by the toxic metabolites (toxins) produced by pathogenic microorganisms. In some cases, toxic metabolites result from the interaction between host and parasite. From the pathological viewpoint, toxins can be classified into host-specific and nonspecific toxins.

a. Host-Specific Toxins. Host-specific toxins (HSTs) are also called host-selective toxins or pathotoxins, and are essential to evoking diseases for the producer pathogens. As described, HSTs show toxicity only to the susceptible plants and susceptible cultivars of a plant species but not to the nonhost plants or resistant cultivars. For pathogens which produce HSTs, the intensity of the pathogenicity is correlated with the amount of toxin productivity. The plant species or cultivars of a species tolerant to HSTs are not affected by the toxin-producing pathogens. Further, all physiological and biochemical phenomena caused by the infection with HST-producing pathogens can be caused by the HSTs.[59] At the present time, at least 14 pathogenic fungi are known to produce HSTs, and almost all HST-producing fungi belong to the genera *Helminthosporium* and *Alternaria* (Table 2).

As to the mechanism on the specificity of HSTs, several hypotheses such as inactivation in resistant plants, inhibition of permeability of toxin to the action site in resistant plants, presence of receptors in susceptible plants, and so on, have been presented. Of these, the inactivation[60] and receptor hypotheses[61] have been the subjects of discussion; however, at the present time indirect data that support the receptor hypothesis have been accumulated. For example, HV toxin (victorin) is an HST produced by the Victoria blight pathogen of oats, *Helminthosporium victoriae*, which is only pathogenic to an oat cultivar, Victoria. Wolpert and Macko[62] isolated and characterized victorin-binding protein from susceptible but not from resistant genotypes of oats by treating both genotypes with radiolabeled victorin. The molecular weight of the victorin-binding protein is about 100 kDa.[62] This 100-kDa protein is also present in resistant oats, but does not bind to victorin. The black spot pathogen of Japanese pear, *Alternaria alternata* Japanese pear pathotype, produces HST named AK toxin. Otani et al.[63] also suggested the presence of AK toxin receptor in the susceptible Japanese pear. That is, the plasma membrane fractions of susceptible and resistant pear fruits were mixed with AK toxin I at concentrations higher than those causing necrosis on susceptible pear leaves; and these mixtures were spotted on susceptible leaves to examine the effect of added membrane fractions. As a result, the toxicity of AK toxin decreased by the addition of the fraction from susceptible fruits as compared with that from resistant ones. Although a nonspecific binding substance is present in both susceptible and resistant pears, this result also supports the receptor hypothesis.

In addition to scientific interests, HSTs can be used as a very useful tool for a breeding program of resistant plants. Wheeler and Luke[64] first applied victorin in breeding of resistant mutants of oats and concluded that correlation between resistance to a parasite and resistance to its toxin is a necessary requisite for such use of toxins. These studies were examples of mass screening of HST-insensitive clones in genetically susceptible plants.

Tobacco plants have not yet been known to have any dominant gene(s) for susceptibility or resistance against the brown spot disease. The pathogenic fungus, *Alternaria alternata* tobacco pathotype, produces HST named AT toxin toxic only to *Nicotiana* plants. Therefore, by using the toxin, Thanutong et al.[65] selected protoplast-derived calli from tobacco leaves that were insensitive to the toxin, and regenerated intact plants from the toxin-insensitive calli. After five generations, all of the progenies (R_5) showed resistance to the toxin and also to the pathogen. Successful application of the in vitro screening method using HSTs as selective materials has been undertaken in some other crops such as the maize plant resistant to *Helminthosporium maydis* race T,[66] the sugarcane plant resistant to eyespot disease caused by *Helminthosporium sacchari*,[67] and so on.

HSTs are thus the primary determinants of pathogenicity, but only 14 species of pathogenic fungi have been found to produce HSTs. Therefore, other plant pathogens must have other factors to determine the pathogenicity, especially in obligate parasites.

b. Nonspecific Toxins. Nonspecific toxins are the metabolites of plant pathogens, which show injurious effects not only to host plants but also to nonhost plants. Although many phytotoxic compounds have been found in culture filtrates of pathogenic microorganisms, there is no guarantee that these compounds are involved in the plant pathogenesis because the composition of culture media and conditions of culture are different from that in host plants. To know the importance of the nonspecific toxin in pathogenesis, we must at least detect the toxin in the diseased plant. Thus, "the toxin produced in the infected host by pathogen and/or its hosts, which function in the production of disease, but is not itself the initial inciting agent of disease", was defined as vivotoxin.[68]

Because these nonspecific toxins cannot explain host specificity, they have been defined as the secondary determinant of disease. Nevertheless, some of these toxins are necessary for a successful infection on the host plants, and others are responsible for the symptom development or decreasing yield.

From much available information on the role of nonspecific toxins, several notable examples will be presented.

The "bakanae" disease of rice is caused by *Gibberella fujikuroi*. Kurosawa[69] found that all the characteristic symptoms — the unusual elongation of rice leaves and internodes, chlorosis, and reduced tillering — caused by pathogenic fungi can be reproduced by treating rice plants with a cell-free culture filtrate on which pathogenic fungus was grown. Two active components were isolated in crystalline form from the culture filtrate and named gibberellin A and B.[70,71] After that active research was conducted by many scientists from all over the world, and they found that gibberellin is the mixture of many related derivatives. At the beginning of this research, gibberellin was considered as a special factor for the symptom development in this disease, but at present it is well-known that gibberellins are growth-regulating hormones widely distributed in the plant kingdom. Other than gibberellins, fusaric acid which inhibits the growth of rice plant was isolated by Yabuta et al.[72] from culture filtrate of the "bakanae fungus", *G. fujikuroi*. Fusaric acid is also produced by many soil-inhabiting plant pathogens belonging to genus *Fusarium* (imperfect stage of *Gibberella*), and known to be a factor causing wilting of many plants. Fusaric acid injures the plasma membrane of various plants and increases the permeability of the cell.[73] As a result, many cell components such as K, Na, Ca, amino acids, and so on are leaked onto the surface of plants, and increases in the osmotic pressure of the leaf surface occur. The water of cell sap exudes out and hence the plant wilts. Furthermore, fusaric acid has a strong chelating activity with metal ions and inhibits the enzymes of the respiratory enzymes of plant cells.[73,74] This is another mechanism showing that this toxin works as the virulence factor.

The *Helminthosporium* leaf spot pathogen of rice plant, *Cochliobolus miyabeanus (Ophiobolus miyabeanus* or *Helminthosporium oryzae)*, is a typical perthophyte, an organism that kills the host cells prior to hyphal penetration. In regard to the perthophytic activity of this fungus, Yoshii[75] suggested the involvement of some toxins as the result of his microscopic observation that a number of cells surrounding the invaded part of the leaf had already been killed prior to hyphal invasion. Later, the antibiotic ophiobolin was isolated from the culture filtrate of *C. miyabeanus* and was shown to be produced by all strains of *C. miyabeanus* tested.[76] The chemical structure of this substance

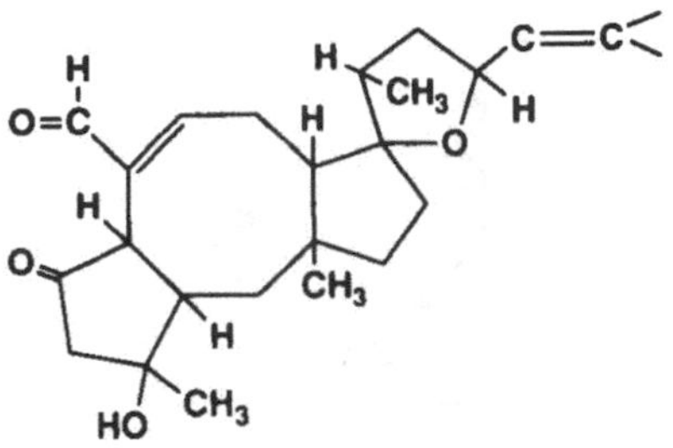

FIGURE 5. Chemical structure of ophiobolin A.

was first determined by Nozoe et al.[77] to be a C_{25} terpenoid having a novel structure. Recently a series of ophiobolin has been characterized,[78] and the original substance is called ophiobolin A (Figure 5). Ophiobolin A was found in the rice leaves inoculated with the pathogenic fungus (vivotoxin) and produces a number of small necrotic spots on rice leaves immersed in the solution of this toxin at a concentration of 10 ppm or more.[79] Other ophiobolins are also reported to cause the characteristic symptom of this disease at 10^{-4}–10^{-5} *M*.[78] Because all physiological changes — such as increase in respiration, inhibition of photosynthesis, protein synthesis, and RNA synthesis — which occur by inoculation with the pathogenic fungus are reproduced by treatment with ophiobolin A, ophiobolin seems to be a causal factor for the physiological symptom of this disease.[80]

Hyperplastic diseases in plants caused by fungal and bacterial pathogens are believed to be due to a high amount of auxin and cytokinin produced by pathogens that perturb the hormone balance in host tissues. Hirata[81] found that a large amount of free auxin is contained in the galls of many plant species caused by species of *Exobasidium, Uromyces, Gimnosporangium, Ustilago, Albugo,* and so on as compared with healthy tissues; and that the amount of free auxin contained in tissues was proportional to the degree of the over growth. Hirata[82] also demonstrated that some of these fungi (except obligate parasitic fungi) produced auxin in culture filtrate, especially in tryptophan-supplemented media. The genus *Taphrina* which causes hyperplasia has also been shown to produce auxin and cytokinin in culture media.[83]

Recently it was clarified that *Taphrina wiesneri, T. derformans,* and *T. pruni* — causal pathogens of cherry witches broom, peach leaf curl, and plum pocket, respectively — synthesize indole acetic acid (IAA) from tryptophan via indole-3-pyruvate and indole-3-acetoaldehyde,[84] differently from those in bacteria in which indole acetoamide is an intermediate[85,86] (Figure 6).

Bacterial pathogens which cause gall and knot on plant tissues also produce auxin and cytokinin, and the full sequence of genes encode enzymes that convert tryptophan to indoleacetic acid had been determined.[87] The *tms-1* gene in *Agrobacterium tumefaciens* and *iaa M* in *Pseudomonas syringae* pv.

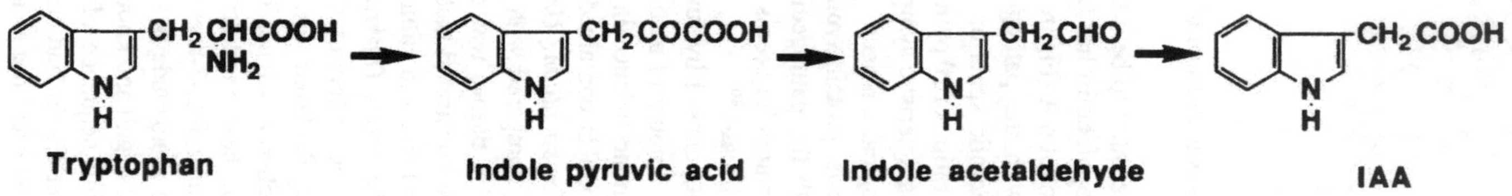

FIGURE 6. Possible pathway of IAA synthesis by genus *Taphrina* from tryptophan.

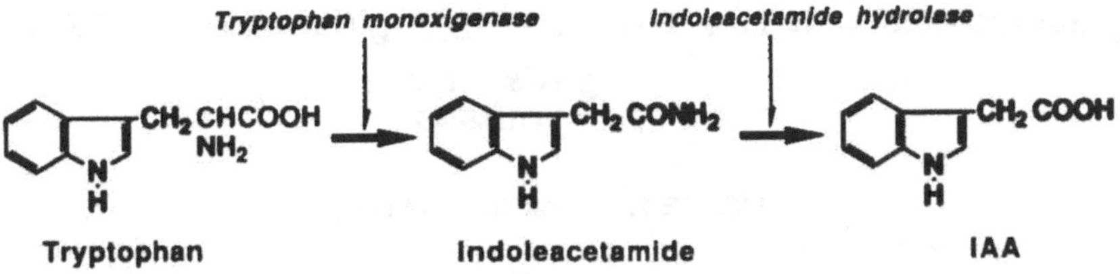

FIGURE 7. Pathway of IAA synthesis by *Agrobacterium tumefaciens* and *Pseudomonas syringae* pv. *savastanoi*.

savastanoi encode tryptophan monooxygenase, and *tms-2* and *iaa H* encode indoleacetamide hydrolase[87] (Figure 7). The Dutch elm disease is caused by *Ceratocystis ulmi* which is transmitted by elm bark beetles as described earlier. The rapid death of the elm tree is due to the wilt toxin, ceratoulmin, produced by the pathogenic fungus.[88] Ceratoulmin is a protein with a molecular weight of 7628 and with 75 amino acid residues, and is detected immunologically in field white elms that were infected naturally by means of vector transmission.[89] Ceratoulmin seems to belong to semi-host-specific toxins because nonhost plants respond to the toxin more mildly than did the host plants. Ceratoulmin enhances the respiration rate and membrane permeability of the host cells. When measured by electrolyte loss from plant cells, a clear-cut host specificity is observed. Namely, nonhost plants, apple, Japanese pear, poplar, tobacco, barley, and tomato are not affected at all, but host plants respond in accordance with the type of response to the pathogen itself.[89]

Fusicoccin is the major nonspecific toxin produced by *Fusicoccum amygdali*, the bud canker pathogen of almond and peach.[90] Before the canker development, wilting and drying of leaf blades occur as the early symptoms. In artificially inoculated trees, fusicoccin concentration in wilting leaf tissue reaches 0.1–0.6 mg/kg fresh weight, biologically active concentration, as determined by a specific radioimmunoassay.[91] As for the mechanism of fusicoccin on the wilting of leaves, the stomatal aperture is enhanced and the transpiration is increased.

Other than hyperplasia-inducing pathogens, many bacterial plant pathogens produce toxins that cause hypoplastic symptoms. *Pseudomonas syringae* pv. *tabaci*, the causal bacterium of wildfire disease of tobacco, produces a toxin, tabtoxin. Tabtoxin is a dipeptide, and chemical structure was determined as shown in Figure 8 by Stewart.[92] Production of tabtoxin leads to the symptom of a number of bacterial diseases in many plant species, tobacco, oats, beans,

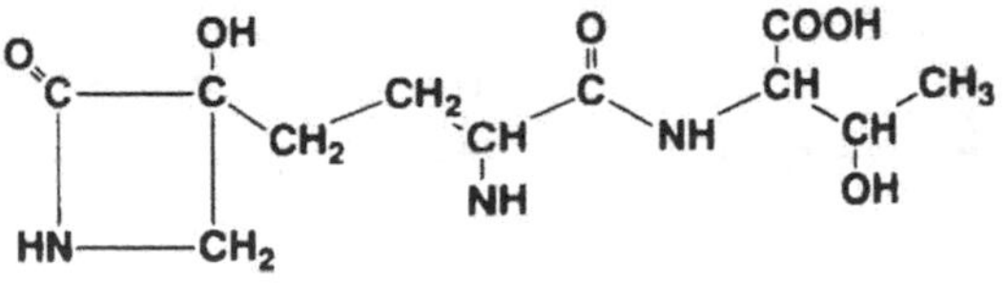

FIGURE 8. Chemical structure of tabtoxin.

$$\text{glutamic acid} + NH_3 + ATP \underset{\textit{glutamine synthase}}{\rightleftharpoons} \text{glutamine} + ADP + Pi$$

FIGURE 9. Glutamine synthase.

soybeans, etc.[93] As the mode of action, Sinden and Durbin[94] clarified that this toxin inhibits glutamine synthase (Figure 9). This was supported by the fact that chlorosis of the leaf caused by tabtoxin could be reversed by L-glutamine. The importance of tabtoxin as the virulence factor will be discussed in a later chapter.

Pseudomonas syringae **pv.** ***phaseolicola*** is the causal agent of the halo blight of beans. At the site of infection a green-yellow chloretic zone is formed and ornithine accumulates.[95] A nonspecific toxin — phaseolotoxin — that reproduces these symptoms was isolated, and the chemical structure was determined (Figure 10).[96] Phaseolotoxin is a potent inhibitor of ornithine transcarbamylase that synthesizes citrulline from ornithine; and as a result, a large amount of ornithine accumulates around the halo by inoculation with the bacterial pathogen and also by treatment with the toxin.[97]

Coronatine is produced by several pathovars of *Pseudomonas syringae* such as atropurpurea[98] and glycinea.[99] The chemical structure was established by Ichihara et al.[100] as indicated in Figure 11.

Treatment of Italian ryegrass with coronatine showed chlorosis and browning similar to that caused by inoculation with *P. syringae* pv. *atropurpurea*.[98] At present, it is known that the coronatine-producing ability of the bacterium is associated with 58 MDa plasmid, pCOR1. That is, all of the pCOR1-cured strain of *P. syringae* pv. *atropurpurea* lost the ability to produce coronatine,[101] and the cured strains restored coronatine productivity by reintroducing the plasmid.[102]

c. Toxic Substances Resulting from Host-Parasite Interactions. Abnormal metabolites produced as a result of the host-parasite interaction by plants sometimes act as the virulence factor. The later stage of the soft rot of persimmon fruit caused by anthracnose fungus is due to the tissue macerating enzymes of the host origin as described before. This conclusion came from the careful experimental results obtained by Tani.[47] He compared fungal

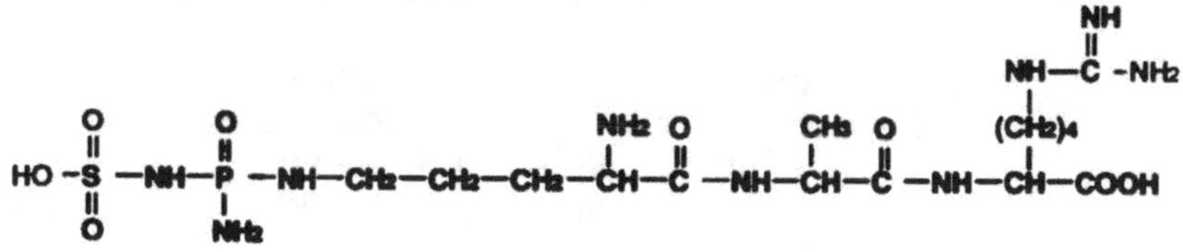

FIGURE 10. Chemical structure of phaseolotoxin.

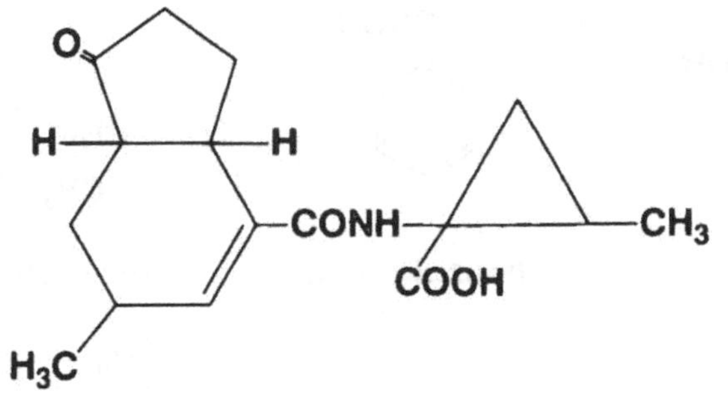

FIGURE 11. Chemical structure of coronatine.

pectolytic enzymes and macerating enzymes produced in the uninfected softened persimmon tissues and found that the later stage of pathogenesis is caused by enzymes of host origin.[46] The fact that the treatment of persimmon fruit with 2,4-D induces the same enzymes also supports this conclusion.[47]

In the powdery mildew disease of peas, the colonies formed on pea leaves develop without limitation and finally the infected leaves wither. The withering is not responsible for the toxins of pathogenic fungus, *Erysiphe pisi*, but for the phytoalexin, pisatin, which is produced as the defense substance of the host. When the concentration of pisatin reaches 300 ppm (fresh weight basis), the plasma membrane of peas is injured. In contrast, the pathogenic fungus is tolerant to pisatin and unaffected; hence it continues to grow. The cause of increase in respiration in powdery mildew infected peas is also due to the accumulated pisatin which uncouples the oxidative phosphorylation of pea mitochondria.[103] Pisatin was also demonstrated to be toxic to human erythrocytes and an uncoupler of oxidative phosphorylation in mammalian mitochondria.[104]

These facts suggest that the mechanism of injury to the host plant is very complicated in obligate parasitic diseases compared with nonobligate parasitic ones. The direct cause of death of the pea plant by infection with powdery mildew fungus seems to be a kind of "suicide" by the toxic effect of pisatin produced by the pea itself. However, indirectly the virulence factor for this disease may be the stimulus to induce pisatin, the elicitor produced by the pathogenic fungus. Thus, the tolerance of this fungus to pisatin is most likely to contribute to the virulence. If the pathogen is sensitive to the phytoalexin of the host, the development of the fungus may be inhibited, it ceases to produce elicitors, and the level of phytoalexin does not reach the lethal dose.

The rapid wilting of Japanese and some European pines infected by the pinewood nematode *(Bursaphelenchus xylophilus)* is considered to be caused by some toxic compounds, because tissues of infected pine trees are seriously affected before the invasion by nematodes.[105] Later, toxins were isolated from naturally infected pine trees and characterized as catechol,[106] benzoic acid,[106] 8-hydroxycarvotanacetone,[106] dihydroconiferylalcohol,[106,107] and

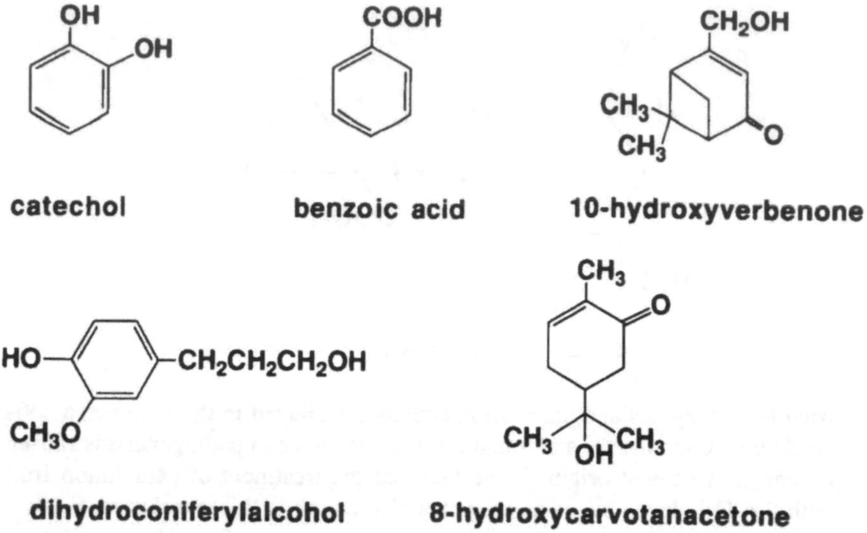

FIGURE 12. Toxic metabolites isolated from pine tree infected by pinewood nematode.

10-hydroxyverbenone[107] (Figure 12). The outline was reviewed by Oku.[108] Of these, catechol seems to be produced by a bacterium associated with the pathogenic nematode because the bacterium produces catechol when incubated in the pine leaf juice.[109] Other toxic metabolites are likely to be abnormal metabolites of the pine tree which are produced by the stimuli of infection by the pinewood nematode. These materials show a high synergistic effect in toxicity. Both benzoic acid and 8-hydroxycarvotanacetone become detectable 20 days after inoculation with the pinewood nematode, when the first symptom — reduction of oleoresin exudation — occurs. The content of benzoic acid reaches more than 300 μg/g of dry wood by 50 days. The other toxins also accumulate at a biologically active level in wood of infected pine standings in the forest.[108,110]

E. Pathogenicity of Plant Pathogenic Viruses

Viruses infect only living cells and cause a variety of physiological disturbances in their host or hosts. This physiological disturbance — disease in our term — comes from the complete host dependency on virus multiplication because viruses have no metabolic function such as a protein synthetic, energy producing system, or multiplication.

In general, viruses are composed of nucleic acid, ribonucleic acid (RNA) or deoxyribonucleic acid (DNA), and the coat protein; and some are covered with the envelope. The multiplication of viruses is performed by the expression of genetic information encoded in the genome of their nucleic acid.

According to Zaitlin and Hull,[111] there are about 630 plant parasitic viruses in 77% of which the nucleic acid is single, positive-stranded RNA; this can synthesize proteins directly as mRNA on the host ribosomes. Almost all important plant pathogenic viruses belong to this group. The other 13% has single, negative-stranded RNA as the genome. These types of viruses can synthesize protein after the formation of complementary RNA by the function of RNA-polymerase present in the virus particles. The remainders are double-stranded RNA and DNA, and single-stranded DNA viruses. To date, the primary structure of more than 10 plant viral genomes and the genes involved in these genomes have been determined. The development of genetic engineering, especially the reverse transcription technique, made possible the analysis of gene function in viral RNA from its cDNA. Further, more than 15 infectious viral RNAs have been synthesized in vitro. The origin of plant viruses are also discussed from the evolutional viewpoint by comparing the structure with that of animal viruses.

1. Entry of Viruses into Plants

Viruses themselves cannot enter into a plant through the cuticle and cell wall. They enter into plants through wounds formed mechanically or biologically. Many plant viruses are mediated by insects, mites, nematodes, and sometimes by fungi. Among these, about 300 plant viruses are mediated by insects. The infection process of a plant virus can be divided into following three steps: (1) absorption, invasion, and uncoating; (2) recognition of host cell and multiplication in the primary infected cell; and (3) transfer to the other cells.

On the first step, many of the mechanisms of absorption and invasion remain unclear compared with the second and third steps. The reason may be due to the low infectivity by artificial inoculation. The injury made by carborundum also affects the understanding of actual interaction between the virus particle and the host cell.

The processes of absorption and invasion of the tobacco mosaic virus (TMV) into the isolated protoplast of tobacco mesophyll have been observed by the electron microscope. Under the presence of poly-L-ornithine, one side of the rod-shaped TMV particle (negatively charged) is absorbed onto the plasma membrane of the protoplast, and then the particles are introduced into the protoplast by way of endocytosis.[112,113] The role of poly-L-ornithine is thought to diminish the negative charge of the protoplast membrane. However, there are other opinions that poly-L-ornithine injures the membrane; and the virus particle is introduced through the injured part, but not by way of endocytosis.[114,115] The positively charged brome mosaic virus (BMV) and pea enation mosaic virus (PEMV) are reported to be absorbed and infect tobacco protoplast without the addition of poly-L-ornithine.[116,117] Thus absorption of the virus particle onto the plasma membrane seems to be due to the electrostatic manner.

Ehara[118] concluded from careful experiments using cowpea tissue and cowpea mosaic virus (CMV) that the virus particle is introduced into the host cell through a mechanical wound of the membrane. The wounding of the cell membrane causes the difference of electric potential from the uninjured membrane. The difference disappears during the curing process of the membrane; and at the same time, the infectivity of CMV into the cowpea cell disappears. Thus, the actual mechanism of entry of a virus particle remains unclear.

After the entrance into a plant cell, the virus particle uncoats the protein from the 5′ end. The naked part of TMV-RNA immediately binds to the ribosomes of the host cell to form striposome, and synthesizes the proteins with molecular weights of 130 and 180 kDa.[119] These proteins function as RNA-dependent RNA replicase to form template RNA, replicative form, and new viral RNAs including subgenome RNAs for coat protein and 30-kDa protein.[120]

The uncoating of TMV and the formation of striposome occur in several resistant plants; BMV uncoats in the cell of the nonhost plant (cabbage), and therefore the recognition for multiplication and the host specificity may be determined at a later stage than uncoating.[121,122]

2. Recognition of Host Cell and Multiplication in the Primary Infected Cell

The host range of viruses in the natural ecosystem sometimes is narrower than that determined by inoculation tests. The host range of insect-mediated viruses may be determined by the host range of vector insects. Thus, the outbreak of a virus disease is, in some parts, determined by the chance of contact between plants and viruses.

How the invading viruses cause disease is largely governed by the response of plant genes to viral gene products. The response of the host to a viral infection is classified into the following four categories by Zaitlin and Hull:[111]

1. Immunity — no infection and multiplication occur even when viruses invade into the plant cell.
2. Subliminal infection — invaded viruses multiply only in the primary infected cell but do not transfer to the other cells; hence, no symptoms appear.
3. Local infection — the viruses infect only in the primary invaded and few surrounding cells; hence, the infection is localized.
4. Susceptibility — invaded viruses multiply and transfer systemically.

a. Immunity. In the case of an immune response, if the functional RNA replicase is not formed, the transcription of genomic RNA does not occur. For example, RNA replicase of TYMV is composed of two subunits, 115-kDa protein encoded in the viral RNA and 45-kDa protein of host origin.[123,124] Even if the 115-kDa protein is synthesized in the turnip yellow mosaic virus plant cell, (TYMV)-RNA cannot be transcripted when the 45-kDa subunit is not present in the cell.

b. Subliminal Infection. This response may be governed by the following two mechanisms: proteins for transfer between host cells, which are translated from the virus genome, are inactivated, or the translation is inhibited.[125,126]

c. Localized Infection. An example of this is TMV-OM and TMV-L producing a systemic mosaic and local necrotic lesions, respectively, on tobacco having an *N'* gene. To know the factor that induces a local lesion, many recombinant viruses between OM and L were produced and the reaction to *N'* tobacco was examined. As a result the factor that induces necrotic local lesion in *N'* tobacco was the coat protein of the L strain.[127]

d. Susceptibility. The mechanism that inhibits the multiplication and transfer to the other cells is not present in the host cells, or some mechanism that enhances the multiplication of virus is present in host cells. In general, resistance is a rule and susceptibility may be the exception in plant-virus interactions, because each virus has its own host or hosts and cannot infect the majority of plant species.

A brief explanation of the multiplication of viruses in the primary invaded cell is described earlier in Section IV and is therefore omitted in this section.

3. Translocation of Virus Particles Within Host Plants

The translocation of virus particles within host plants had been believed to be a passive phenomenon. However, it has been clarified that the gene for transfer among the host cells is coded in the virus genome from the analytical experiment of the genome structure of the temperature-sensitive TMV mutant.

The Ls 1 strain, a mutant of the TMV-L strain, cannot translocate from the infected host cell to the neighboring healthy cells at a higher temperature.[128] That is, when the leaves of tobacco having an *N* gene are inoculated with the L or Ls 1 strains and kept at 22 or 32°C, the L strain multiplies and transfers to the neighboring cells. In the L strain the lesion enlarges even at 32°C, but in Ls 1 it does so only at 22°C. That is, the increase in the number of infected cells in the tomato does not increase at 32°C when inoculated with Ls 1. However, no difference was found in the multiplication of the Ls 1 strain in the tomato protoplast between 22 and 32°C. From these experimental results it was concluded that the Ls 1 strain is a temperature-sensitive mutant that cannot transfer between host cells.

Then three kinds of proteins synthesized in an *in vitro* protein synthetic system with L and Ls 1 were compared, and it was found that only the 30-kDa protein was different.[129] The comparison of nucleotide sequences encoding the 30-kDa protein of both strains clarified that the mutation is the exchange of the 153th proline (from N terminal) to serine in the Ls 1 strain.[130] This mutation occurs only at the 1 nucleotide exchange. Thus the point mutation of the nucleotide sequence causes the change of function of TMV.

This phenomenon has been confirmed by using many artificial mutants.[131] The localized presence of a 30-kDa protein in plasmodesmata before the accumulation of TMV in the host cell was observed under the electron microscope.[132]

REFERENCES

1. Agrios, G. N., *Plant Pathology*, Academic Press, New York, 1978, 172 and 435.
2. Ogura, H., Studies on saprophytic behaviour of soil born pathogenic fungi. V. Fungal succession on plant debris in soil, *Resp. Rep. Kochi Univ.*, 17, 13, 1968.
3. Nelson, R. R., The evolution of parasitic fitness, in *Plant Disease*, Horsfall, J. G. and Cowling, E. B., Eds., Academic Press, New York, 1979, 23.
4. Oku, H., Determinant for pathogenicity without apparent phytotoxicity in plant diseases, *Proc. Jpn. Acad.*, 56 (Ser. B), 367, 1980.
5. Wheeler, B. E. J., Fungal parasites of plants, in *The Fungi III*, Einsworth, G. C. and Sussman, A. F., Eds., Academic Press, New York, 1968, 179.
6. Flor, H. H., Host-parasite interaction in flax rust — Its genetics and other implications, *Phytopathology*, 45, 680, 1955.
7. Hiura, U., Hybridization between varieties of *Erysiphe graminis*, *Phytopathology*, 52, 664, 1962.
8. Oku, T., Yamashita, S., Doi, Y., and Hiura, U., Genetic analysis of resistance of what cultivars to races and some *formae speciales* of *Erysiphe graminis* DC, *Ann. Phytopathol. Soc. Jpn.*, 52, 700, 1986.
9. Flor, H. H., The complementary genetic systems in flax and flax rust, *Adv. Genet.*, 8, 29, 1956.
10. Westcott, C., *Plant Disease Handbook*, D. Van Nostrand, Toronto, 1950, 3.
11. Horsfall, J. G. and Dimond, A. E., The plant disease, in *Plant Pathology I*, Horsfall, J. G. and Dimond, A. E., Eds., Academic Press, New York, 1959, 7.
12. Agrios, G. N., *Plant Pathology*, Academic Press, New York, 1978, 4.
13. Wagener, W. W. and Davidson, R. W., Heart rot in living trees, *Bot. Rev.*, 20, 61, 1954.
14. Marsh, R. W., Observation on apple canker. II. Experiment on the incidence and control of shoot infection, *Ann. Appl. Biol.*, 26, 458, 1939.
15. Foister, C. E., Wilson, A. R., and Boyd, A. E. W., Dry-rot disease of the potato. I. Effect of commercial handling methods on the incidence of the disease, *Ann. Appl. Biol.*, 39, 29, 1952.
16. Peace, T. R., *Pathology of Trees and Shrubs*, Oxford University Press, London, 1962, 753.
17. Mamiya, Y. and Enda, N., Transmission of *Bursaphelenchus lignicolus* (Nematoda:Aphelenchoidae) by *Monochamus alternatus* (Coleoptera:Cerambycidae), *Nematologica*, 18, 159, 1972.
18. Wheeler, B. E. J., Fungal parasite of plants, in *The Fungi III*, Ainsworth, G. C. and Sussman, A. S., Eds., Academic Press, New York, 1968, 179.
19. Brown, W., The physiology of host parasite relations, *Bot. Rev.*, 2, 236, 1936.
20. von Ramm, C., Histological studies of infection by *Alternaria longipes* on tobacco, *Phytopathol. Z.*, 45, 391, 1962.

21. Balls, W. L., Infection of plants by rust fungi, *New Phytol.*, 4, 18, 1905.
22. Dickinson, S., Studies in the physiology of obligate parasitism. II. The behaviour of the germ tubes of certain rusts in contact with various membranes, *Ann. Bot.*, 13, 213, 1949.
23. Hurd-Karrer, A. M. and Rodenheiser, H. A., Structure corresponding to appressoria and substomatal vesicles produced on nutrient-solution agar by cereal rusts, *Am. J. Bot.*, 34, 377, 1947.
24. Pristou, R. and Gallegy, M. E., Leaf penetration by *Phytophthora infestans*, *Phytopathology*, 44, 81, 1954.
25. Wiltshire, S. P., Infection and immunity studies on the apple and pear scab fungi *(Venturia inaqualis* and *V. pirina)*, *Ann. Appl. Biol.*, 1, 335, 1915.
26. Corner, E. J. H., Observation on resistance to powdery mildews, *New Phytol.*, 34, 180, 1985.
27. Sakamoto, M., Studies on the resistance of rice to *Pyricularia oryzae*, in *Jubilee Publication in Commemoration of 60th Birthday of Professor M. Sakamoto*, Suzuki, N., Takahashi, Y., Tomiyama, K., Ui, T., Tsuyama, H., and Kikumoto, T., Eds., Publication Committee Tohoku University, Sendai, 1968, 1.
28. Oku, H., Histochemical studies on the infection process of *Helminthosporium* leaf spot disease of rice plant with special reference to disease resistance, *Phytopathol. Z.*, 44, 39, 1962.
29. Wood, R. S. K., Chemical ability to breach the host barriers, in *Plant Pathology II*, Horsfall, J. G. and Dimond, A. E., Eds., Academic Press, New York, 1960, 233.
30. Dickinson, S., The mechanical ability to breach the host barriers, in *Plant Pathology II*, Horsfall, J. G. and Dimond, A. E., Eds., Academic Press, New York, 1960, 203.
31. Oku, H. and Sumi, H., Mode of action of pentachlorobenzyl alcohol, a rice blast control agent. Inhibition of hyphal penetration of *Pyricularia oryzae* through artificial membrane, *Ann. Phytopathol. Soc. Jpn.*, 34, 250, 1968.
32. Shirane, N., Analysis of Factors Related to the Infection of Host by the Genus, *Botrytis*, Ph.D. thesis, submitted to Okayama University, 1989, 20.
33. Melander, L. W. and Craigie, J. H., The nature of resistance of *Beberis* spp. to *Puccinia graminis*, *Phytopathology*, 17, 45, 1927.
34. Oku, H., unpublished data.
35. Kolattukudy, P. E., Enzymatic penetration of the plant cuticle by fungal pathogens, *Ann. Rev. Plant Pathol.*, 23, 233, 1985.
36. Kolattukudy, P. E. and Crawford, M. S., The role of polymer degrading enzymes in fungal pathogens, in *Molecular Determinant of Plant Diseases*, Nishimura, S., Vance, C. P., and Doke, N., Eds., Japan Scientific Society Press, Tokyo/Springer-Verlag, Berlin, 1987, 75.
37. Brown, W., Studies in the physiology of parasitism. VIII. On the exosmosis of nutrient substances from host tissue into infection drop, *Ann. Bot.*, 36, 101, 1922.
38. Kerr, A. and Flentje, N. T., Host infection in *Pellicularia filamentosa* controlled by chemical stimuli, *Nature (London)*, 179, 204, 1957.
39. Akai, S., Histology of defense in plants, in *Plant Pathology I*, Horsfall, J. G. and Dimond, A. E., Eds., Academic Press, New York, 1959, 302.
40. Betts, C. C. V., Observation on the infection of wheat by loose smut (*Ustilago tritici* [Pers.] Rostr.), *Br. Mycol. Soc. Trans.*, 38, 465, 1955.

41. Kitajima, H. and Kajiwara, T., Blossom blight of apple, in *Crop Diseases with Colored Photos*, Yokendo, Tokyo, 1961–1962.
42. Wescott, C., *Plant Disease Handbook*, D. Van Nostrand, New York and London, 1950, 65.
43. Dickson, J. G., *Diseases of Field Crops*, McGraw-Hill, New York, 1959, 174.
44. Yoshii, H., Patho-histologic observations on tomato inoculated with *Pyricuralia oryzae*, the rice-blast fungus, *Ann. Phytopathol. Soc. Jpn.*, 18, 14, 1948.
45. Griffiths, G. W. and Beck, S. D., Effect of antibiotics on intercellular symbiotes in the pea aphid, *Acyrthosiphon pisum*, *Cell Tissue Res.*, 148, 287, 1974.
46. Tani, T., Studies on the phytopathological physiology of Kaki anthracnose, with special reference to the role of pectic enzymes in the symptom development on Kaki fruits, *Mem. Fac. Agric. Kagawa Univ.*, 18, 1, 1965.
47. Tani, T., The relation of soft rot caused by pathogenic fungi to pectic enzyme production by the host, in *The Dynamic Role of Molecular Constituents in Plant-Parasite Interaction*, Mirocha, C. J. and Uritani, I., Eds., The American Phytopathological Society, St. Paul, MN, 1967, 40.
48. Starr, M. P. and Moran, F., Eliminative spirit of pectic substances by phytopathogenic soft-rot bacteria, *Science*, 135, 920, 1962.
49. Tribe, H. T., Studies in the physiology of parasitism. XIX. On the killing of plant cells by enzymes from *Botrytis cinerea* and *Bacterium aroideae*, *Ann. Bot.*, 19, 351, 1955.
50. Echandi, E., Van Gundy, S. D., and Walker, J. C., Pectolytic enzymes secreted by soft-rot bacteria, *Phytopathology*, 47, 549, 1957.
51. Basham, H. G. and Bateman, D. F., Killing of plant cells by pectic enzymes: the lack of direct injurious interaction between pectic enzymes or their soluble reaction products and plant cell, *Phytopathology*, 65, 141, 1975.
52. Garibaldi, A. and Bateman, D. F., Pectic enzymes produced by *Erwinia chrysanthemi* and their effect on plant tissue, *Physiol. Plant Pathol.*, 1, 25, 1971.
53. Ried, J. L. and Collmer, A., Construction and characterization of *Erwinia chrysanthemi* mutant with directed deletion in all of the pectate lyase structural genes, *Mol. Plant-Microbe Interact.*, 1, 32, 1988.
54. Goto, M. and Okabe, N., Studies on pectin-methyl-esterase secreted by *Erwinia carotovora* (Jones) Holland, with special reference to the cultural conditions relating to the enzyme production and the difference of enzyme activity due to the kind of strains, *Ann. Phytopathol. Soc. Jpn.*, 27, 1, 1962.
55. Konishi, K., Inoue, Y., and Higuchi, T., Decomposition of lignin by *Coriolus versicolor*. IV. Effect of laccase type enzyme on the inter phenylpropane linkage of lignin, *Mokuzai Gakkaishi*, 18, 571, 1972.
56. Oku, H., Biochemical studies on *Cochliobolus miyabeanus*. III. Some oxidizing enzymes of rice plant and its parasites and their contribution to the formation of the lesions, *Ann. Phytopathol. Soc. Jpn.*, 23, 169, 1957.
57. Oku, H., Biochemical studies on *Cochliobolus miyabeanus*. IV. Fungicidal action of polyphenols and the role of polyphenoloxidase of the fungus, *Phytopathol. Z.*, 38, 341, 1960.
58. Oku, H., Role of parasite enzymes and toxins in development of characteristic symptoms in plant disease, in *The Dynamic Role of Molecular Constituents in Plant Parasite Interaction*, Mirocha, C. J. and Uritani, I., Eds., American Phytopathological Society, St. Paul, MN, 1967, 239.
59. Pringle, R. B. and Scheffer, R. P., Host-specific toxins, *Annu. Rev. Phytopathol.*, 2, 133, 1964.

60. Romanko, R. R., Physiological basis for resistance of oat to Victora blight, *Phytopathology*, 49, 32, 1959.
61. Scheffer, R. P. and Pringle, R. B., Uptake of *Helminthosporium victoriae* toxin by oat tissue, *Phytopathology*, 54, 832, 1964.
62. Wolpert, T. J. and Macko, V., Victorin binding to proteins in susceptible and resistant oat genotypes, in *Phytotoxins and Plant Pathogenesis*, Graniti, A., Durbin, R. D., and Ballio, A., Eds., Springer-Verlag, Berlin, 1989, 39.
63. Otani, H., Kohmoto, K., and Nishimura, S., Action sites for AK-toxin produced by Japanese pear pathotype of *Alternaria alternata*, in *Host-Specific Toxins: Recognition and Specificity in Plant Disease*, Kohomoto, K. and Durbin, R. D., Eds., Tottori University Press, Tottori, Japan, 1989, 107.
64. Wheeler, H. E. and Luke, H. H., Mass screening for disease resistant mutants in oats, *Science*, 122, 1229, 1955.
65. Thanutong, P., Furusawa, I., and Yamamoto, M., Resistant tobacco plants from protoplast-derived calli selected for their resistance to pathotoxin, *Theor. Appl. Genet.*, 66, 209, 1983.
66. Gegenbach, B. G. and Green, C. E., Selection of T-cytoplasm maize callus cultures resistant to *Helminthosporium maydis* race T pathotoxin, *Crop Sci.*, 15, 645, 1975.
67. Heinz, D. J., Sugarcane improvement through induced mutations using vegetative propagules and cell culture techniques, in *Induced Mutation in Vegetatively Propagated Plants*, International Atomic Energy Agency, Vienna, 1973, 53.
68. Dimond, A. E. and Waggoner, P. E., On the nature and role of vivotoxins in plant disease, *Phytopathology*, 43, 229, 1953.
69. Kurosawa, E., Experimental studies on the secretion of *Fusarium heterosporum* on rice plants, *J. Nat. Hist. Soc. Formosa*, 16, 213, 1926.
70. Yabuta, T. and Sumiki, Y., Isolation of gibberellin, *J. Agric. Chem. Soc. Jpn.*, 14, 1526, 1938.
71. Yabuta, T. and Hayashi, T., Biochemical studies on "bakanae fungus" of rice. II. Isolation of gibberellin, the active principle which produces slender rice seedlings, *J. Agric. Chem. Soc. Jpn.*, 15, 257, 1939.
72. Yabuta, T., Kambe, K., and Hayashi, T., Biochemistry of the bakanae-fungus. I. Fusaric acid, a new product of the bakanae fungus, *J. Agric. Chem. Soc. Jpn.*, 10, 1059, 1934.
73. Broun, R., Uber wirkungsweise und Umwandlungen der Fusarinsäure, *Phytopathol. Z.*, 39, 197, 1960.
74. Tamari, K. and Kaji, J., Studies on the mechanism of injurious action of fusaric acid on plant growth, *J. Agric. Chem. Soc. Jpn.*, 27, 144, 147, 159, 1952–1953.
75. Yoshii, H., On the mode of infection of *Helminthosporium oryzae* (*Ophiobolus miyabeanus* Ito et Kurib.) to rice plant, *Ann. Phytopathol. Soc. Jpn.*, 9, 170, 1939.
76. Nakamura, M. and Ishibashi, K., On the new antibiotic "ophiobolin" produced by *Ophiobolus miyabeanus*, *J. Agric. Chem. Soc. Jpn.*, 32, 732, 1958.
77. Nozoe, S., Morisaki, M., Tsuda, K., Iitaka, Y., Takahashi, N., Tamura, S., Ishibashi, K., and Shirasaka, M., The structure of ophiobolin, C_{25}-terpenoid having a novel skeleton, *J. Am. Chem. Soc.*, 87, 4968, 1965.

78. Sugawara, F., Takahashi, N., Yun, C., Strobel, G., and Clardy, J., The phytotoxic ophiobolins produced by *Drechslera oryzae*, their structures and biological activities on rice, *Abstr. 5th ICPP*, 1988, 221.
79. Nakamura, M. and Oku, H., Biochemical studies on *Cochliobolus miyabeanus*. IX. Detection of ophiobolin in the diseased rice leaves and its toxicity against higher plants, *Ann. Takamine Lab.*, 112, 266, 1960.
80. Chattopadhyay, A. K. and Sammadder, K. R., Comparative physiological changes induced by *Helminthosporium oryzae* infection and ophiobolin, *Phytopathol. Z.*, 98, 118, 1980.
81. Hirata, S., Studies on the phytohormone in the malformed portion of the diseased plants. III. Auxin formation on the culture grown *Exobasidium, Taphrina and Ustilago* spp., *Ann. Phytopathol. Soc. Jpn.*, 22, 153, 1957.
82. Hirata, S., Studies on the phytohormone in the malformed portion of the diseased plants. I. The relation between growth ratio and the amount of free auxin in fungus galls and virus infected plants, *Ann. Phytopathol. Soc. Jpn.*, 19, 33, 1954.
83. Kern, H. and Naef-Roth, S., Zur Bildung von Auxinen und Cytokininen durch *Taphrina*-Arten, *Phytopathol. Z.*, 83, 193, 1975.
84. Yamada, T., Tsukamoto, H., Shiraishi, T., Nomura, T., and Oku, H., Detection of indoleacetic acid biosynthesis in some species of *Taphrina* causing hyperplastic diseases in plants, *Ann. Phytopathol. Soc. Jpn.*, 56, 532, 1990.
85. Kosuge, T. and Yamada, T., Virulence determinants in plant pathogen interactions, in *Molecular Determinant of Plant Diseases*, Nishimura, S., Vance, C. P., and Doke, N., Eds., Japan Scientific Society Press, Tokyo/Springer-Verlag, Berlin, 1987, 171.
86. Van Onckelen, H., Prinsen, E., Inze, D., Rudelsheim, P., Van Lijsebettens, M., Follin, A., Shell, J., Van Montagu, M., and Degreef, J., *Agrobacterium* T-DNA gene 1 codes for triptophane 2-monooxigenase activity in tobacco crown gall cells, *FEBS Lett.*, 198, 357, 1986.
87. Yamada, T., Palm, C. J., Brooks, B., and Kosuge, T., Nucleotide sequence of the *Pseudomonas savastanoi* indoleacetic acid genes show homology with *Agrobacterium tumefaciens* T-DNA, *Proc. Natl. Acad. Sci. U.S.A.*, 82, 6522, 1985.
88. Takai, S., Richard, W. C., and Stevenson, K. J., Evidence for the involvement of cerato-ulmin, the *Ceratocystis ulmi* toxin in the development of Dutch elm disease, *Physiol. Plant Pathol.*, 23, 257, 1983.
89. Takai, S., Host-specific factor in Dutch elm disease, in *Host-Specific Toxins*, Kohmoto, K. and Durbin, R. D., Eds., Tottori University Press, Tottori, Japan, 1989, 75.
90. Graniti, A., The role of toxins in the pathogenesis of infections by *Fusicoccum amygdali* Del. on almond and peach, in *Host Parasite Relations in Plant Pathology*, Kiraly, Z. and Ubrizsy, G., Eds., Research Institute of Plant Protection, Budapest, 1964, 211.
91. Graniti, A., Fusicoccin and stomatal transpiration, in *Host-Specific Toxins*, Kohmoto, K. and Durbin, R. D., Eds., Tottori University Press, Tottori, Japan, 1989, 143.
92. Stewart, W. W., Isolation and proof of structure of wild fire toxin, *Nature (London)*, 229, 174, 1971.
93. Ballio, A., Structure-activity relationships, in *Toxins in Plant Disease*, Durbin, R. D., Ed., Academic Press, New York, 1981, 259.

94. Sinden, S. L. and Durbin, R. D., Glutamine synthetase inhibition: possible mode of action of wild fire toxin from *Pseudomonas tabaci*, *Nature (London)*, 219, 379, 1968.
95. Patil, S. S., Toxin production by pathogenic bacteria, *Annu. Rev. Phytopathol.*, 12, 259, 1974.
96. Mitchell, R. E., Isolation and structure of a chlorosis inducing toxin of *Pseudomonas phaseolicola*, *Phytochemistry*, 15, 1941, 1976.
97. Patil, S. S., Kolattukudy, P. F., and Dimond, A. E., Inhibition of ornithine carbamoyltransferase from bean plant by the toxin of *Pseudomonas phaseolicola*, *Plant Physiol.*, 46, 752, 1970.
98. Nishiyama, K., Sakai, R., Ezuka, A., Ichihara, A., Shiraishi, K., and Sakamura, S., Phytotoxic effect of coronatine in halo blight lesions of Italian ryegrass, *Ann. Phytopathol. Soc. Jpn.*, 42, 219, 1977.
99. Mitchell, R. E. and Young, H., Identification of chlorosis-inducing toxin of *Pseudomonas glycinea* as coronatine, *Phytochemistry*, 17, 2028, 1978.
100. Ichihara, A., Shiraishi, K., Sato, H., Sakamura, S., Nishiyama, K., Sakai, R., Furusaki, A., and Matsumoto, T., The structure of coronatine, *J. Am. Chem. Soc.*, 99, 636, 1977.
101. Sato, M., Nishiyama, K., and Shirata, A., Involvement of plasmid DNA in the productivity of coronatine by *Pseudomonas syringae* pv. *atropurpurea*, *Ann. Phytopathol. Soc. Jpn.*, 49, 522, 1983.
102. Sato, M., *In planta* transfer of the gene(s) for virulence between isolates of *Pseudomonas syringae* pv. *atropurpurea*, *Ann. Phytopathol. Soc. Jpn.*, 54, 20, 1988.
103. Shiraishi, T., Oku, H., Isono, M., and Ouchi, S., The injurious effect of pisatin on the plasma membrane of pea, *Plant Cell Physiol.*, 16, 939, 1975.
104. Oku, H., Ouchi, S., Shiraishi, T., Utsumi, K., and Seno, S., The toxicity of a phytoalexin, pisatin, to mammalian cells, *Proc. Jpn. Acad.*, 52, 33, 1976.
105. Mamiya, Y., Inoculation of the first year pine seedlings with *Bursaphelenchus lignicolus* and the histopathology of diseased seedling, *J. Jpn. For. Soc.*, 62, 176, 1980.
106. Ueda, T., Oku, H., Tomita, K., Sato, K., and Shiraishi, T., Isolation, identification, and bioassay of toxic compounds from pine tree naturally infected by pine wood nematode, *Ann. Phytopathol. Soc. Jpn.*, 50, 116, 1984.
107. Bolla, R. I., Shaheen, F., and Winter, R. E. K., Phytotoxins in *Bursaphelenchus xylophilus* induced pine wilt, in *The Resistance Mechanisms of Pine Against Pine Wilt Disease*, Dropkin, V., Ed., University of Missouri Press, Columbia, 1984, 119.
108. Oku, H., Role of phytotoxins in pine wilt disease, *J. Nematol.*, 20, 245, 1988.
109. Oku, H., Shiraishi, T., and Ouchi, S., Possible participation of toxins in pine wilt disease, in *Proc. XVII IUFRO World Congr.*, Kyoto, 1981, 281.
110. Oku, H., Yamamoto, H., Ohta, H., and Shiraishi, T., Effect of abnormal metabolites from nematode-infected pine on pine seedling and pine wood nematodes, *Ann. Phytopathol. Soc. Jpn.*, 51, 303, 1985.
111. Zaitlin, M. and Hull, R., Plant virus host interactions, *Annu. Rev. Plant Physiol.*, 38, 291, 1987.
112. Cocking, E. C., An electron microscopic study of the initial stage of infection of isolated tomato fruit protoplast by tobacco mosaic virus, *Planta*, 68, 206, 1966.

113. Otsuki, Y., Takebe, I., Honda, Y., and Matusi, C., Ultrastructure of infection of tobacco mesophyll protoplasts by tobacco mosaic virus, *Virology*, 49, 188, 1972.
114. Burges, J., Motoyoshi, F., and Fleming, E. N., Effect of poly-L-ornithine on isolated tobacco mesophyll protoplast: evidence against stimulated pinocytosis, *Planta*, 111, 199, 1973.
115. Burges, J., Motoyoshi, F., and Fleming, E. N., The mechanism of infection of plant protoplasts by viruses, *Planta*, 112, 323, 1973.
116. Motoyoshi, F. and Hull, R., The infection of tobacco protoplasts with pea enation mosaic virus, *J. Gen. Virol.*, 24, 89, 1974.
117. Motoyoshi, F., Bancroft, J. B., and Watts, J. W., The infection of tobacco protoplasts with a variant of brome mosaic virus, *J. Gen. Virol.*, 25, 31, 1974.
118. Ehara, Y., Molecular mechanism of host response to viruses, in *Recent Advances in Physiological Plant Pathology*, Oku, H., Kohmoto, K., Doke, N., Odani, H., Tsuge, T., and Kodama, K., Eds., The Publishing Committee of Recent Advances in Physiological Plant Pathology, Nagoya, 1991, 195.
119. Wilson, T. M. A., Cotranslational disassembly of tobacco mosaic virus in vitro, *Virology*, 137, 255, 1984.
120. Goelet, P., Lomonossoff, G. P., Buttler, P. J. G., Akam, M. E., Gait, M. J., and Karn, J., Nucleotide sequence of tobacco mosaic virus RNA, *Proc. Natl. Acad. Sci. U.S.A.*, 79, 5818, 1982.
121. Kiho, Y., Machida, H., and Oshima, N., Mechanisms determining the host specificity of tobacco mosaic virus. I. Formation of polysomes containing infecting viral genome in various plants, *Jpn. J. Microbiol.*, 16, 109, 1969.
122. Shaw, J. G., In vitro removal of protein from tobacco mosaic virus after inoculation of tobacco leaves. II. Some characteristics of the reaction, *Virology*, 37, 109, 1969.
123. Candresse, T., Mouches, T., and Bove, J. M., Characterization of the virus encoded subunit of turnip yellow mosaic RNA replicase, *Virology*, 152, 322, 1986.
124. Mouches, T., Candresse, T., and Bove, J. M., Turnip yellow mosaic virus RNA-replicase contains host and virus-encoded subunits, *Virology*, 134, 78, 1984.
125. Cheo, P. C., Subliminal infection of cotton by tobacco mosaic virus, *Phytopathology*, 60, 41, 1970.
126. Salzinskin, M. A. and Zaitlin, M., Tobacco mosaic virus replication in resistant and susceptible plants: in some resistant species virus is confined to a small number of initially infected cells, *Virology*, 121, 12, 1982.
127. Saito, T., Meshi, T., Takamatsu, N., and Okada, Y., Coat protein gene sequence of tobacco mosaic virus encodes a host response determinant, *Proc. Natl. Acad. Sci. U.S.A.*, 84, 6074, 1997.
128. Nishiguchi, M., Motoyoshi, F., and Oshima, N., Behaviour of a temperature sensitive strain of tobacco mosaic virus in tomato leaves and protoplast, *J. Gen. Virol.*, 39, 53, 1978.
129. Leonard, D. A. and Zaitlin, M., A temperature-sensitive strain of tobacco mosaic virus determined in cell-to-cell movement generates an altered viral-coded protein, *Virology*, 117, 416, 1982.
130. Ohno, T., Takamatsu, N., Meshi, T., Okada, Y., Nishiguchi, M., and Kiho, Y., Single amino acid substitution in 30 K protein of TMV defective in virus transport function, *Virology*, 131, 255, 1983.

131. Meshi, T., Watanabe, Y., Saito, T., Sugimoto, A., Maeda, T., and Okada, Y., Function of the 30 Kd protein of tobacco mosaic virus: involvement in cell-to-cell movement and dispensability for replication, *EMBO J.*, 6, 2557, 1987.
132. Tomenius, K., Claphan, D., and Meshi, T., Localization by immuno gold cytochemistry of virus coded 30 K protein in plasmodesmata of leaves infected with tobacco mosaic virus, *Virology*, 160, 363, 1987.

CHAPTER 2

Resistance of Plants Against Pathogens

I. RESISTANCE IS THE RULE AND SUSCEPTIBILITY IS THE EXCEPTION

In general, living beings including microorganisms tend to reject other species. This rejection reaction might be established during the evolutional process to maintain the purity of species.

Biologists use this property for the classification of species. That is, they judge the species as the same when fertilization occurs and can maintain stable offspring. Animals show very strict mechanisms to keep the purity of species from the first step of fertilization. The sperm produces the lytic enzyme for the membrane of the egg cell, but the substrate specificity of the enzyme is very strict and only acts on the egg membrane of the same species. If the sperm could get into the egg cell of the other species, it could not fertilize it because of the difference in the number and quality of the chromosomes.

Although the fertilization occurs when the male and female belong to the same species, the compatibility at the organ, tissue, and cell level is more strict, especially in highly evolved living beings like mammals. The success or failure of the transplantation of organs in medical treatment is solely dependent on how the rejection reaction can be prevented. In the case of higher plants, grafting can only be successful between the same or related species.

The resistance of plants and animals against microbial attack seems to be the same type of rejection reaction. That is, living beings, as a rule, reject invading microorganisms. Thus, the pathogens of plants and animals are limited in number; and in the case of vertebrates, the individual that is invaded by a pathogen usually is never attacked by the same pathogen. However, these vertebrates become a very good source of nutrients for microorganisms and allow them to multiply after their death.

The same is true in higher plants. According to the list "Common Names of Economic Plant Diseases in Japan" published by the Phytopathological Society of Japan, the fungal pathogens of the potato plant have only 20 species. However, the boiled extract of the potato tuber is one of the best culture media and can grow thousands of fungal species. This simple fact suggests that the potato tuber contains all the necessary nutrients for fungi, but other than 20 species most cannot parasitize on the living potato because of the rejection reaction (resistance response). Thus, in living beings the resistance is the rule, and pathogens have to overcome the resistance of their hosts to establish infection.

Just as the potato pathogen cannot invade the other species of plants, the blast fungus of rice cannot infect the living potato. Thus, pathogens have their own host or hosts which they infect and derive nutrients from. This phenomenon is called the "host specificity" of pathogens, and the ability to overcome the resistance and parasitize on host plants might be acquired under coexistence with the hosts during the evolutional process. Plants have also acquired the strategies to oppose these pathogens.

The resistance of plants against pathogens can be classified into two types according to the mechanism, static resistance and dynamic resistance. Static resistance is also called preformed or constitutional resistance. This type of resistance is dependent on the constitutional characteristics of the normal, uninfected plants. Dynamic resistance is sometimes called active defense or induced resistance, which is expressed after the microbial attack.

II. STATIC RESISTANCE EQUALS CONSTITUTIONAL RESISTANCE

A. Physical Barrier Against Microbial Penetration

1. Strength of Plant Surface

Many pathogenic fungi do not make use of either wounds or natural openings but penetrate the unbroken surface of plants to enter into plants. This type of penetration is called direct penetration or cuticular penetration as described in Chapter 1. For these pathogenic fungi, the physical properties of the plant surface such as hardness and thickness affect the penetration; and a lot of experimental evidence has been accumulated.

The species of *Berberis* which is not susceptible to *Puccinia graminis* was reported to have a thick wall, even in young leaves.[1] Sakamoto[2] examined the relationship between the resistance of rice plant against blast disease caused by *Pyricularia oryzae* and the resistance to needle puncture, and found positive correlations except in cultivars having vertical resistance genes to this disease. For example, among the same cultivar or variety of rice plant, cultural conditions such as excess application of nitrogen fertilizer, low content of soil water, low temperature, etc. decrease the resistance of the rice plant. Rice plants cultivated in such conditions decrease the resistance to needle puncture. For the rice blast fungus and the *Helminthosporium* leaf spot fungus, the motor cells and the guard cells of stomata are the entry sites through which fungi penetrate most easily. More than half of the total invasion is through motor cells. For this reason it was considered that lignification of the outer wall of the motor cell does not occur rapidly and remains as the pectocellulosic condition for a long time, while most other epidermal cell walls are cellulosic and lignified sooner.[3]

It is well-known that the silicate content of leaves is positively correlated to the resistance of rice against the blast and the *Helminthosporium* leaf spot fungi. Under suitable conditions, silicate deposits in the surface layer of the motor cell and increases the resistance to fungal penetration.[4] Thus the resistance to penetration plays an important role in disease resistance.

2. Stomatal Barrier as Resistance Factor

Many pathogenic fungi and bacteria enter into plants through stomata. A plausible idea that closed stomata might prevent the entry of fungal germ tubes is not always supported by experimental evidence; that is, some fungi cannot enter plants when stomata are closed, but some can. Therefore, it is difficult to draw a general conclusion about the resistance from the behavior of stomata.

Hart[5] suggested that the uredospore germ tube of *Puccinia graminis* could not penetrate through the closed stomata, but could enter only when the stomata were open. Therefore, some cultivars in which stomata open late in the morning are resistant to rust disease, because the germ tube dies by desiccation before the stomata open. Hart named this type of resistance "functional resistance". The germ tube of a sugar beet pathogen, *Cercospora beticola,* cannot penetrate through stomata under conditions resulting in closed stomata.[6]

On the other hand, there are many observations that the closed stomata offer no barrier to the penetration. In leaf rust of wheat caused by *Puccinia triticina,* the formation of an appressorium over an open stoma often causes prompt closure of the stoma and the penetration hypha emerging from the appressorium forces entry through closed stomata.[7] The stomatal activity also is reported to play a small role in the prevention of the infection of *Pinus strobus* by *Cronartium ribicola.*[8]

Thus the role of stomatal behavior as the barrier for fungal penetration varies in each host-parasite combination. It is likely that the entry is made directly by the germ tube (no formation of appressorium) as in *C. beticola,* and closed stomata may contribute a barrier. For the type of entry shown by rust uredospores, in which the penetration hypha from appressorium forced entry into the closed stomata, the closed stomata may not be a barrier. In other words, this type of penetration seems to be more effective than the type in which the germ tube enters directly through stomata.

B. Components of Plants as Nutrients for Pathogens

After the infection has been established, plant pathogens grow and multiply on and in host plants by consuming components of host plants as nutrients. When the components are suitable for the nutrients, the disease outbreak may be severe. On the contrary, the plant in which components are unsuitable for the pathogen may be resistant. Many experimental results on this point have been accumulated in Japan on diseases of the rice plant because rice is the most important crop in Asian countries. Review papers on *Helminthosporium* leaf spot,[9] *Pellicularia* sheath-blight,[10] stem rot,[11] and blast[12] are available. According to these data, there are positive correlations between the susceptibility to these diseases and the availability of components of rice plants as the source of nutrients for these pathogens. For example, there are many reports that pathogenic fungi grow better in the media containing components from susceptible rice cultivars than those from resistant ones. It is well-known that excess manuring of nitrogen fertilizer caused severe outbreaks of these diseases. In such plants with contents of free amino acids, especially aspartic and glutamic acid which are the best nutrients for pathogenic fungi, diseases have been found to increase.

Notwithstanding these experimental results, it seems unlikely that the resistant plants contain nutrients for pathogens so small in amount as to suppress the growth of pathogens. This is experienced in the artificial culture of pathogens. That is, the small difference of the amount of components in culture media does not seriously affect the growth of microorganisms. Further, Sakamoto[2] pointed out that in the leaf sheath of the rice plant fertilized by excess nitrogen, the inoculated hyphae grow into neighboring cells very rapidly, just as the defense reactions are suppressed.

Thus, the increase in susceptibility by high manuring of nitrogen does not seem to be only the nutritional problem, but other effects such as decrease in cellular function due to the excess elongation, and softening the tissues also may lower the resistance of the host.

C. Antimicrobial Components of Plants

Plants containing antimicrobial components may be resistant to some diseases of which pathogens are sensitive to the components. Ingham[13] classified these antimicrobial plant components into the following four groups:

1. Prohibitins — preinfectional plant metabolites which can markedly reduce or completely halt the in vivo development of an organism unadapted to its effect

2. Inhibitins — preinfectional plant metabolites which, although present in detectable quantities in apparently healthy plants, must undergo a marked postinfectional increase in their toxic potential is to be fully expressed
3. Postinhibitins — antimicrobial metabolites produced by plants in response to infection (or mechanical or chemical damage) but whose formation does not involve the elaboration of biosynthetic pathways within the tissues of the host
4. Phytoalexins — antibiotics formed in plants via a metabolic sequence induced either biologically or in response to chemical or environmental factors

In this section the role of compounds belonging to groups of prohibitin and postinhibitin is discussed, and inhibitins and phytoalexins will be discussed later in Section III on dynamic resistance. Regarding compounds belonging to prohibitin and postinhibitin, phenolic compounds, glycosides, saponins, etc. have been studied extensively.

1. Phenolic Components

Phenolic compounds — such as chlorogenic acid, tannic acid, coumarin, and the glycoside of phenolic compounds that are contained in many plant species — have more or less antibiotic activity; hence much attention has been paid to these compounds as the cause of disease resistance.

According to many experimental results, the content of phenolics in plants increases after the infection. This will be described later in Section III on active defense.

However, in some cases, the constitutive phenolics play a decisive role in disease resistance.

The onion cultivars that have dried red outer scale are resistant to smudge caused by *Colletotrichum circinans*, but the cultivars having noncolored scale are susceptible. Pioneer work by Link et al.[14,15] and Walker and Link[16] confirmed that the red onion scale contains a sufficient amount of protocatechuic acid[14,15] and catechol[16] to inhibit completely the spore germination of this pathogen and thus accounts for the mature red onions resistance to this disease (Figure 1). These substances are not contained in inner scales of the red onion.

The content of chlorogenic acid in potato tubers was found to be higher in resistant cultivars than in susceptible ones against scab disease caused by

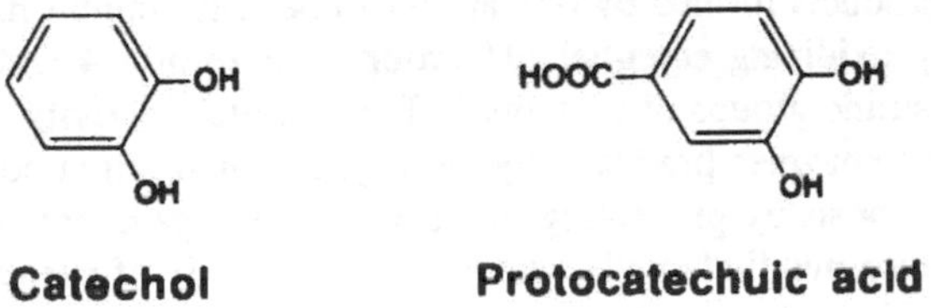

FIGURE 1. Antimicrobial phenolics contained in the red outer scale of onion.

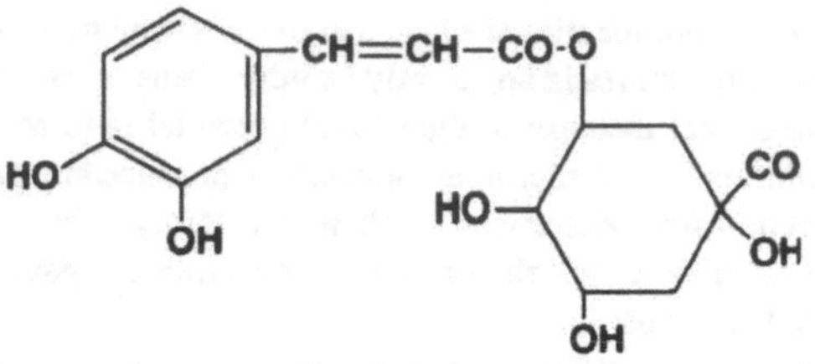

FIGURE 2. Chlorogenic acid, responsible for resistance of potato to scab disease.

Streptomyces scabies[17] (Figure 2). This compound is confined to the outermost tissues where the pathogen multiplies and the tissue around the lenticel, the infection sites.

Polyphenol derivatives such as chlorogenic acid, caffeic acid, flavons, etc. are widely distributed in the plant kingdom. Because polyphenol derivatives are easily oxidized by oxidative enzymes contained in plants or microorganisms, it has been difficult to estimate the exact antibiotic activity of the phenolic substances themselves. I found that antifungal activity of the oxidation product of catechol is very much higher than catechol itself.[18] Namely, the addition of reducing agents such as ascorbic acid greatly reduces the antifungal activity of catechol. This is due to the immediate reduction of the oxidized product of catechol, *o*-quinone, into the original catechol. The presence of an appropriate concentration of polyphenoloxidase inhibitor, diethyldithiocarbamate, also reduces the antifungal activity of catechol against *Cochliobolus miyabeanus* which has very active polyphenoloxidase. This is certainly due to the suppression of formation of the oxidized product of catechol. All these results show that the oxidized product has stronger antifungal activity than the parent catechol.

Because the *Helminthosporium* leaf spot pathogen, *C. miyabeanus*, has especially high polyphenoloxidase activity, the fungus is very sensitive to polyphenols.[19] This might be the reason why the symptom of this disease, necrotic spots, remain small. That is, this is due to the high polyphenoloxidase activity of the pathogenic fungus oxidizing the polyphenol components of rice leaves into more toxic oxidation products; hence the growth of the fungus is inhibited. In other words, a kind of "self-inhibition" of pathogenic fungus seems to occur in rice leaves because of the high polyphenol oxidase activity.

The mechanism of antimicrobial activity of polyphenols has been shown to be that products formed by oxidation of phenolics inhibit many important enzymes by oxidizing essential SH groups[20] or form 1,4 addition products with some amino groups of proteins.[21] The inhibitory activity of polyphenols on pectolytic enzymes produced by pathogenic fungi might contribute to the resistance of hosts by preventing the growth of the pathogen within the host tissue, because middle lamellae are composed mainly of pectic substances.[22] The role of phenolics in resistance was reviewed by Kosuge.[23]

2. Glycosides

Glycosides are widely distributed in the plant kingdom. Some of them are found to hydrolyze into sugar moiety and aglycon by the stimuli of infection, and antibiotic aglycon inhibits the growth of the pathogen. Amygdalin, a glycoside contained in *Rosaceae*, is hydrolyzed to benzaldehyde and cyanide through purnacin and D-mandelonitril by the stimuli of fungal infection, or the mechanical wounding[24] (Figure 3). The role of amygdalin in disease resistance has not been fully elucidated but cyanide, the degradation product, is reported to inhibit the growth of pathogenic fungi.

The content of a phenolglucoside, phlorodzin, contained in apple leaves, does not correlate the resistance to *Venturia inaqualis*.[25] However, in leaves of resistant apple, the cells are degraded by a hypersensitive reaction response to invasion of the pathogenic fungus; and β-glucosidase released from the degraded host cell hydrolyzes the glucoside to phloretin, and then is oxidized by host or microbial oxidases to yield a toxic oxidation product[25] (Figure 4).

These types of inhibitors belong to the postinhibitin in Ingram's term, and it seems better to discuss them in Section III on dynamic resistance.

3. Organosulfur Compounds

Alleiaceous plants contain thioether and produce an uncomfortable odor by wounding. This odor comes from diallylsulfide, the end product of degradation of a component of the alleiaceous plant, alliin (Figure 5). Relative compounds of alliin are also contained in cruciferous plants but the alliin degrading enzyme, alliin lyase, is present only in alleiaceous plants. The allicin, an intermediate of alliin degradation, has antibiotic activity to some fungi and bacteria. However, the pathogenic fungus of garlic, *Penicillium corymbiferum*, is tolerant to allicin; and that is why this fungus is pathogenic to garlic. This pathogenic fungus utilizes allicin as a nutrient, and the growth of hyphae and sporulation are promoted. The growth of nonpathogenic *Penicillium* is inhibited at concentrations higher than 186 μg/mL, but garlic contains 5–20 mg/g fresh weight of allicin, and thus prevents the invasion by many microorganisms.[26,27]

4. Saponins

Saponins are also contained in many plant species, and have pronounced antifungal properties; therefore some of them are believed to play roles in disease resistance. Tomatine in tomatoes, solanine and chaconine in potatoes and avenacin in oats have been studied extensively (Figure 6).

As to the mechanism of the antibiotic activity, it is considered that saponins denature sterols present as a liquid crystal in the plasma membrane forming an insoluble compound; as a result, cell contents leak out to surrounding

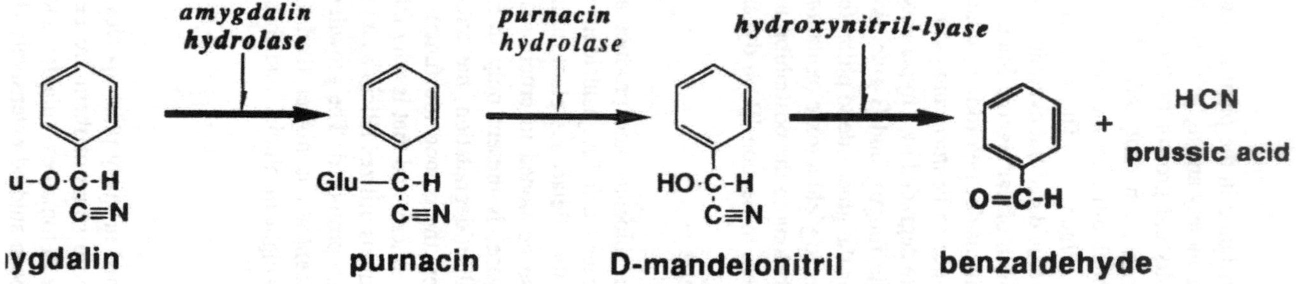

FIGURE 3. Production of cyanide from amygdalin by enzymatic degradation in Rosaceae.

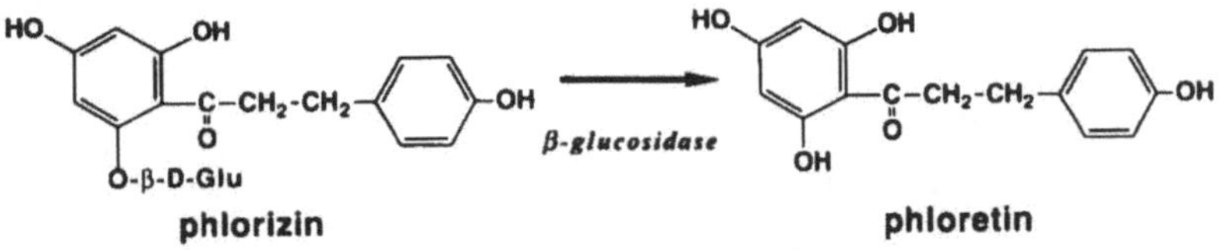

FIGURE 4. β-Glucosidase activity in hypersensitively responding apple tissue may involve the resistance to *Venturia inaqualis* (hydrolysis of phlorizin to phloretin).

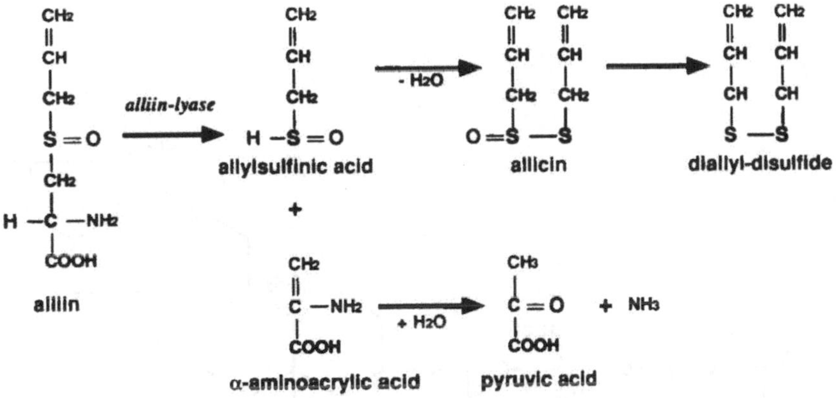

FIGURE 5. Degradation of alliin in an alleiaceous plant by wounding.

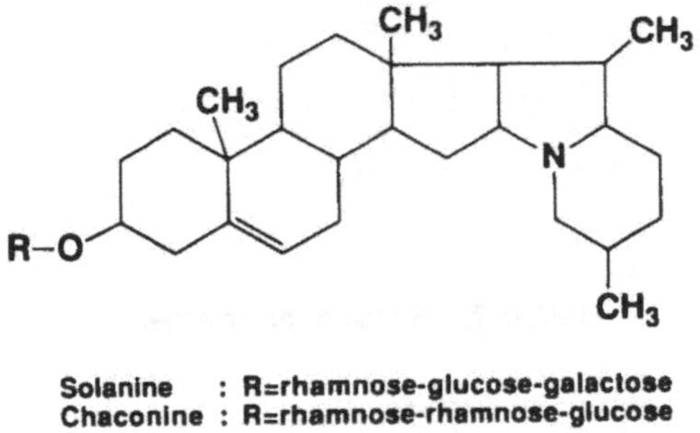

FIGURE 6. Structure of solanine and chaconine in potato.

medium.[28] In general, the pathogenic fungi of the tomato are more tolerant to tomatine than nonpathogens[29] (Figure 7). For example, *Septoria lycopersici* produces a tomatine-inactivating enzyme.[30] The growth of *Helminthosporium carbonum*, a nonpathogen of the potato, is inhibited by solanine and chaconine; but the potato pathogen, *Phytophthora infestans*, is not inhibited by these saponins. This is due to the lack of membrane sterol, the action site of

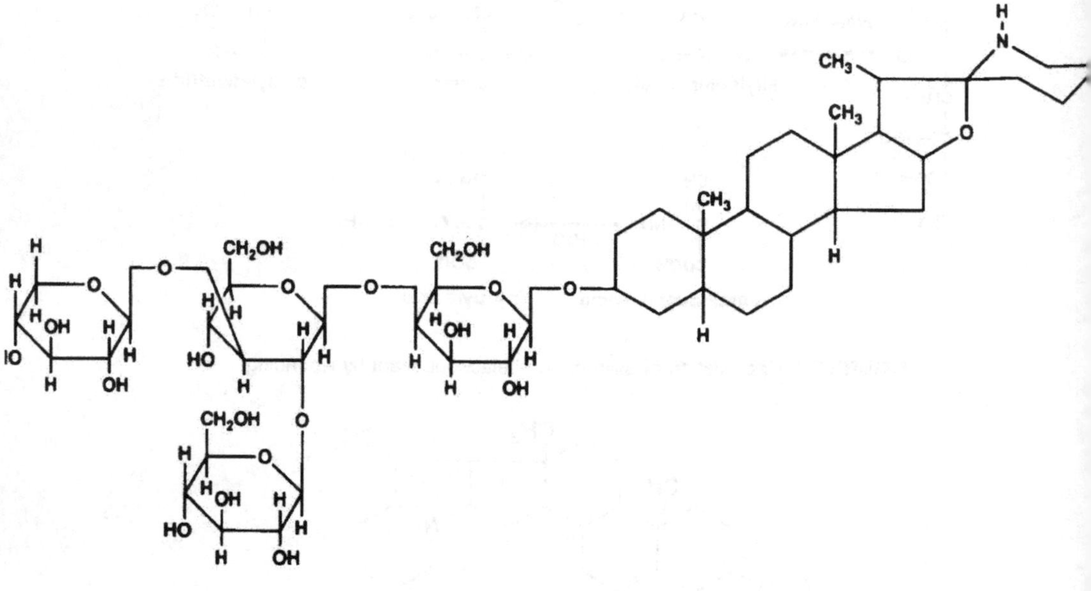

FIGURE 7. Structure of tomatine.

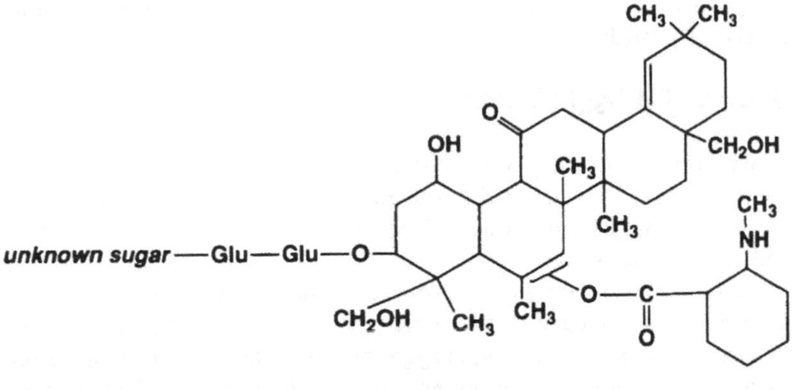

FIGURE 8. Structure of avenacin.

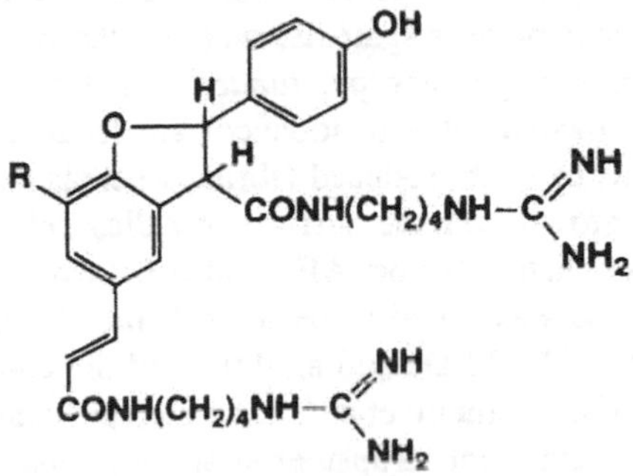

Hordatin A : R=H
Hordatin B : R=OCH3

FIGURE 9. Structure of hordatins.

saponins.[31] Avenacin (Figure 8) contained in oats inhibits the growth of many fungi and bacteria, but does not inhibit *Ophiobolus graminis* f. sp. *avenae* — an oat pathogen — because the pathogen has an avenacin-degrading enzyme — avenacinase — which eliminates the terminal sugar and inactivates it.[32]

5. Other Components

Barley seedlings are resistant to *Helminthosporium sativum* for 5 days or so after germination, but become susceptible thereafter when Ca^{2+} and Mg^{2+} accumulate. These phenomena are explained by the fact that the antifungal compounds, hordatin A and B, in young seedling are inactivated by these cations[33,34] (Figure 9). This was confirmed by the experiment that the addition of chelators of these cations kept the antifungal activity of hordatins. Hordatins strongly inhibit the germination of spores not only of *H. sativum*, but also of many other fungal spores; however, the inhibitory activity to mycelial growth is relatively weak.[34]

D. Lectinlike Agglutinating Factors

Similar to the hemagglutinin contained in leguminous seeds, many plants contain substances which agglutinate bacteria. These substances are also called lectins, and have been acknowledged as recognition factors between plants and bacterial pathogens.

According to the electron microscopic observation, an avirulent strain of the fire blight pathogen of the apple, extracellular polysaccharide deficient mutant of *Erwinia amylovora*, is aggregated in xylem vessels of apple petiole tissue.[35] The virulent strains are not aggregated. The same phenomenon was reported in tobacco. That is, an incompatible pathogen of tobacco, *Pseudomonas syringae* pv. *pisi*, is enveloped near tobacco mesophyll cells by granules, fibrils, and membrane fragments; whereas the compatible pathogen of tobacco, *Pseudomonas syringae* pv. *tabaci*, is not.[36] Similarly, cells of *P. siringae* pv. *pisi* (incompatible to tobacco) are localized on the surface of tobacco callus cells by host-produced fibrils, whereas the compatible *P. syringae* pv. *tabaci* grows over the surface of callus cells.[37] Romeiro et al.[38] extracted two agglutination factors AF I and AF II from apple seed, leaf, and stem tissues.[38] AF II was found to be a small protein of which the relative molecular weight is 12,600 Da and agglutinated the capsule-deficient strain of *E. amylovora*. Then Romeiro et al.[39] examined the component which reacts with AF II from an avirulent, acapsuler strain of *E. amylovora*, and isolated lipopolysaccharide (LPS). While the virulent strain having a capsule was also agglutinated by AF II, AF II was more active to the avirulent strain. Mild acid hydrolysis of the crude LPS gave a milky suspension that was divided into a pellet and supernatant by centrifugation. The supernatant fraction contained all of the receptors for AF. They concluded that the agglutination of *E. amylovora* by AF is due to the charge-to-charge interaction between a highly positively charged protein and a negatively charged component of the bacterial surface.

Sequeira and Graham[40] collected 55 virulent isolates and 34 avirulent isolates of *Pseudomonas solanacearum* from all over the world, and examined their ability to bind to purified potato lectin. As a result, they found that all of the avirulent isolates agglutinated strongly with the lectin, whereas virulent isolates failed to aggregate or aggregated only weakly at much higher lectin concentrations. Inability of virulent strains to bind to lectin was correlated with the presence of extracellular polysaccharide (EPS) which is present in virulent, but not in avirulent isolates. Removal of EPS by repeated washing and centrifugation makes the virulent isolates agglutinate strongly with potato lectin. Addition of EPS to avirulent isolates or washed virulent isolates prevents the agglutination. From these results, they concluded that the EPS of virulent strains prevents the direct binding to potato lectin. Because the binding of lectin to an avirulent cell could be inhibited by chitin oligomers, the receptor is considered to be an N-acetyl glucosamine residue in the bacterial cell wall. Similarly, virulent strains of *Erwinia amylovora* are surrounded by EPS, but avirulent strains of the same bacterium are not.

Thus, the lectin seems to play a role in resistance of plants to bacterial pathogens, and a strategy of pathogens is to have EPS around the bacterial wall.

There are several reports that lectinlike agglutinating factors also play roles in resistance against fungal pathogens.

Uritani and Kojima,[41] using many strains of *Ceratocystis fimbriata* and host plants, reported that the spores of the fungus agglutinate with extracts from nonhost plants and do not agglutinate with that from host plants. For example, spores of the sweet potato strain are not agglutinated with the sweet potato extracts, but spores of the other six nonpathogenic strains are agglutinated within 5 hr. Thus, there is a negative correlation in compatibility between spore agglutination and host-parasite combinations in the sweet potato-*C. fimbriata* system. Such correlations are also found when the other four plants were used, with three exceptions. From these results, they suggested that agglutination factors act as the recognition factor between host and fungal parasites; and further, the resultant agglutination reaction may trigger the forthcoming resistant reaction. The agglutination factor from the sweet potato root appears to be a high molecular weight substance containing protein and polysaccharide moieties, since it is not filtered through the Diaflo membrane and is inactivated partially by pornase and completely by macerozyme. Several other mechanisms of lectin in resistance against fungal disease have been reported. Hohl et al.,[42] through their experimental results using *Phytophthora megasperma* f. sp. *glycinea* and soybean protoplasts, considered that adhesion between host and parasite seems to be essential for compatibility. That is, hyphae of the fungus produce the material which is essential for adhesion, while soybean protoplasts often react to hyphal contact with production of fibrillar polysaccharide materials including glucose-binding and galactose-binding soybean lectins. Kano et al.[43] also isolated partially purified lectin from rice leaves which strongly agglutinates conidia of the blast fungus or the fungal protoplasts. The lectin is considered to participate in the reception of fungal proteoglucomannan onto the host cells. Garas and Kuć[44] isolated partially purified lectins from four potato cultivars with different genes for resistance to the late blight pathogen of potato, *Phytophthora infestans*, and found that all lectins lysed zoospores from five races of the fungus; that is, the lytic ability was nonspecific in all race-cultivar combinations tested. The lytic ability was prevented by trimers and tetramers of N-acetyl glucosamine. These lectins only lyse zoospores but do not agglutinate or lyse cystospores, sporangia, or mycelia. They found that potato lectin precipitated elicitors for terpenoid (phytoalexins of potato), which were extracted from the fungus, although the precipitated lectin-elicitor complex still retained most of the elicitor activity. From these results, they suggested that elicitors may not be exposed on the surface of cystospores, sporangia, and mycelia. Andreu and Daleo[45] obtained two kinds of purified lectins from different parts of a potato tuber, the stem and rose end. Both lectins agglutinate red blood cells, but the specific activity of lectin from the stem end of tuber was 30 times more active than that from the rose end. Both lectins have β-glucosidase activity and hydrolyze β-glucan. Interestingly, the enzymatic activity of the lectin preparation from the rose end was 4–10 times higher than that from the stem end of tubers. Since under certain conditions β-glucan secreted by the fungus was reported to suppress the phytoalexin accumulation in the potato, they consider that these lectins play a role in the resistance of the potato by degrading such suppressor glucan.

III. DYNAMIC RESISTANCE EQUALS ACTIVE DEFENSE EQUALS INDUCED RESISTANCE

Plants defend themselves from the attack of microorganisms by means of a series of events. All these resistance measures expressed after the microbial attack are called dynamic resistance, active defense, or induced resistance; and play a key role in disease resistance.

In vertebrates, it is well-known that the immune reactions, such as phagocytosis by white blood cells or antigen-antibody reaction, protect individuals from invading microbes.

Although the details of defense mechanisms are different between vertebrates and higher plants, the active defense plays a key role in both organisms against diseases.

In the term of "induced resistance", we usually consider two types of phenomena, local induced resistance and systemic induced resistance. Although much of the mechanism on systemic induced resistance remains obscure, this phenomenon is already used as a practical control measure in some plant diseases and will be described in detail in Chapter 5.

A. Local Induced Resistance

1. Papilla Formation

It is well-known that a nipplelike protuberance is formed at an early stage of fungal penetration between the cell wall and plasma membrane. The protuberance further develops and appears to cover the invading hypha when it elongates. The first record of this phenomena is made by De Bary,[47] and since 1863 these structures have received much attention and have been studied as one of the resistance mechanisms in many host-parasite combinations. These structures are called papilla, collar, callosity, lignitubers, or appositional wall thickening; and research on these structures has been reviewed.[48,49]

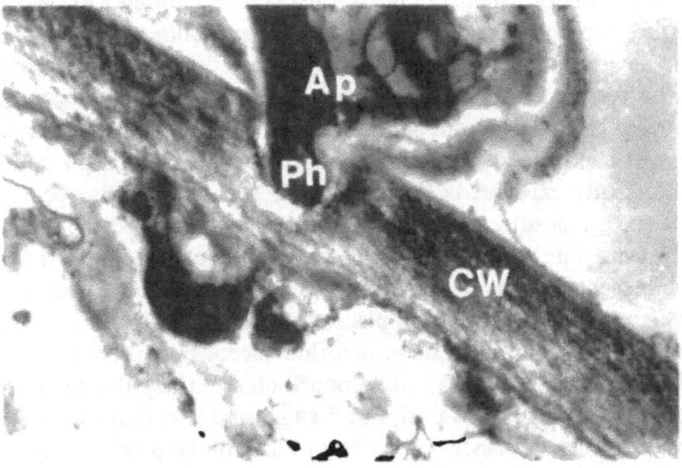

FIGURE 10. Papilla formed on barley epidermis infected by powdery mildew fungus. Ap: Appressorium of powdery mildew fungus; Ph: penetration hypha; CW: host cell wall; Pa: papilla.

Figure 10 shows the papilla formed on the cell wall of barley infected with powdery mildew fungus, *Erysiphe graminis* f. sp. *hordei*.

The importance of papilla in the resistance to fungal penetration seems to be different in each host-parasite combination or varies with environmental conditions. Papillae formed in several plant species do not appear to be infection barriers; while in some plants, papillae seem to effectively prevent the penetration of pathogens.

The results of electron microscopic observation clarified that the wall thickening begins immediately after the formation of appressoria of the pathogenic fungi on the surface of epidermal layers. The papilla-like structures are formed not only by fungal penetration but also by the infection with viruses, bacteria, and nematodes. Since the same structures are formed by injuring cells with a needle or treatment with surface active substances, papillae seem to be formed by the physical stimulus of invading pathogens.[48]

Formerly, it was considered that papillae are formed by deposition of cell wall components which are solubilized by enzymes secreted by invading pathogens. However, according to electron microscopic observation, an attacked cell wall itself is not greatly affected and the components of papillae are different from that of the cell wall; hence at present, papillae are considered to be formed by deposition of cytoplasmic components between cell walls and plasma membranes. Lignin, suberin, cellulose, pectin, gummy substance, silica, callose 1,3-glucan, and peroxidase have been reported to be components of papillae; however, the constituents seem to vary in each host-parasite combination, and also the infection varies even in the same species of a plant.

As to the role of papillae in disease resistance, Stumm and Gessler[50] conducted an experiment using the cucumber and *Colletotrichum lagenarium* system. As the resistant cucumber, they induced systemic resistance by inoculating the first fully expanded true leaf with conidial suspension of *C. lagenarium*. They inoculated both resistance induced and noninduced cucumber leaves, and the penetration ratio of the fungus through papillae was chronologically calculated. As a result, 86% of the papillae penetrated the control (noninduced) cucumber but only 11% penetrated the resistance-induced leaves. Further, they found that where the papillae are not penetrated, the fluorescent reaction with diethanolstilbene-aniline blue was very strong; and where the penetration takes place, fluorescence was very weak. Kita et al.[51] also reported that in barley, autofluorescent papillae were never penetrated by *Erysiphe graminis*, while nonfluorescent papillae were frequently penetrated. Mayama and Shishiyama[52] suggested that fluorescein substances are antimicrobial substances which accumulate in the papillae. Thus, the role of papillae as the resistance barrier against penetration seems to be due to the components of papillae and some antimicrobial substances accumulated in this structure.

2. Lignification of Host Cell Wall

Lignification of the mesophyll cell wall has been known to be caused by wounding or infection by pathogens. Because lignin is an extremely stable substance and plays a role as the barrier to fungal penetration, lignification is considered as one of the defense reactions. As described, lignin can be detected as a component of papillae.

Many examinations hitherto conducted on lignification were done by cytochemical staining with phloroglucinol-HCl, but Asada and Matsumoto[53] extracted, isolated, and chemically examined both lignins present in healthy and diseased radish roots inoculated with *Peronospora parasitica*. According to the results of elemental analysis, they proposed an empirical formula of lignin extracted from a healthy radish root as $C_9H_{10.30}O_{2.20}(OCH_3)_{1.16}$ and that from a diseased root as $C_9H_{8.33}O_{2.80}(OCH_3)_{0.75}$. The healthy lignin is similar to the lignin of a broad-leaved tree which contains syringyl-skeleton derived from sinapyl alcohol as the precursor. Diseased lignin contains less $-OCH_3$ radicals and is similar to needle-leaved tree lignin derived from coumaryl alcohol and coniferyl alcohol as precursors (Figure 11).

These lignins are produced by the oxidation and polymerization of precursors by peroxidase. Asada et al.[54] attribute the difference of healthy and diseased lignin to the difference of substrate specificity of peroxidase isozymes formed in healthy and diseased radish roots.

The lignification of mesophyll cell walls occurs both in susceptible and resistant radish roots when they are infected. However, the lignification occurs very rapidly after the infection in resistant radish roots and serves as a barrier

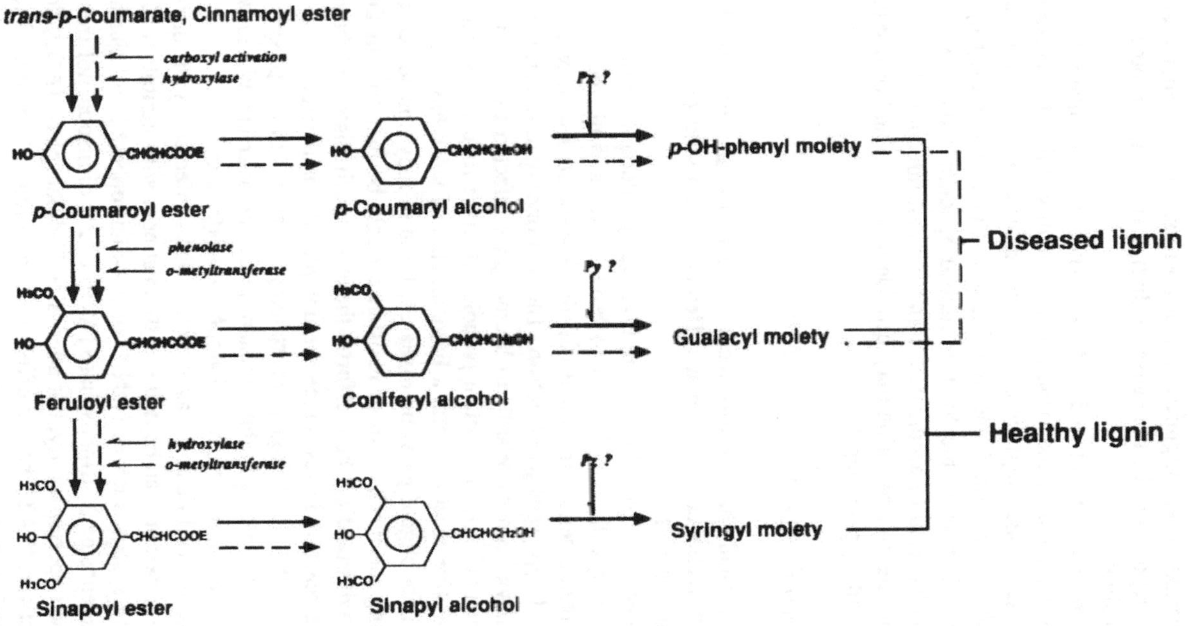

FIGURE 11. Suggested pathway for lignin biosynthesis in healthy (full lines) and diseased (broken lines) plants. Px, Py, Pz: Peroxidase isoenzymes x, y, z. (From Asada, Y. and Matsumoto, I., *Phytopathol. Z.*, 73, 208, 1972. With permission.)

for further growth of the fungus. On the contrary, in the susceptible radish, the lignification occurs behind the expanded fungal hyphae and thus does not serve as a barrier. The mechanism of lignification in downy mildewed radish root will be described later. Lignification of host tissues after microbial infection has been reported to be the cause of disease resistance in other host-parasite combinations.

One of the mechanisms of induced resistance in cucumbers[55,56] and muskmelons[57] is proved to be due to the enhanced ability to lignify in response to infection by pathogenic fungi. In the leaf rust disease of wheat caused by *Puccinia recondita* f. sp. *tritici*, the lignification of cells undergoing hypersensitive response is responsible for the cultivar specific resistance.[58] The stomatal guard cells seem to accumulate lignin by rust infection.[58]

Thus, the rapid lignin deposition may provide a physical and/or chemical barrier to the invading pathogen.[59]

3. Hypersensitive Reaction

Many morphological changes also occur in the protoplasm of host plants by infection with pathogens, and great differences are found between susceptible and resistant plants.

In general, when the resistant plants are invaded by pathogens, the invaded and few surrounding cells rapidly turn to brown and die with the invaded pathogen. This phenomenon, the rapid cell death associated with resistance response, is called hypersensitive reaction. The first finding of hypersensitive cell death may have been by Ward[60] in brome grass infected by brown rust fungus. Since then (1902), many plant pathologists have focused their attention on hypersensitive reactions, and a huge amount of work has been done in obligate and nonobligate parasitic diseases, because they considered that the key to solving the mechanism of resistance might be harbored in these phenomena. Notwithstanding these efforts, the exact mechanism and the role of hypersensitive reaction in resistance are not yet fully understood.

Figure 12 shows the chronological morphological events in the process of hypersensitive cell death of a highly resistant potato cultivar inoculated with zoospores of the late blight fungus, *Phytophthora infestans*.[61]

The hypersensitive cell death in aged potato disks infected by an incompatible race of *P. infestans* can be delayed by treatment with inhibitors of an energy-generating system such as NaN_3[63] or 2,4-dinitrophenol.[62,63] Further, the addition of adenosine 5′-triphosphate (ATP) to 2,4-dinitrophenol-treated tissue restores the hypersensitivity, but adenosine 5′-diphosphate (ADP) does not have such activity.[64] ATP has no effect on compatible interactions. From these experimental results, it was concluded that the energy (generation of ATP) is essential for a hypersensitive reaction. Cytocharacin B which inhibits the formation of actin filaments suppresses the aggregation of cytoplasm and hypersensitive reaction of potato tuber disks caused by the incompatible later blight fungus.[65] The SH-binding high molecular weight reagent such as dex-

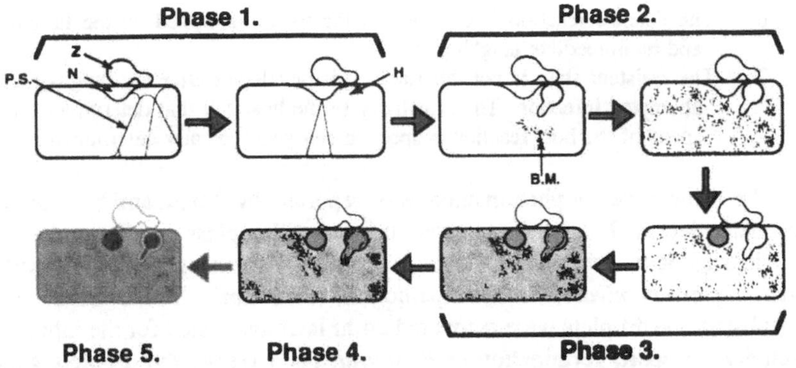

FIGURE 12. Diagram of degeneration progress in the cells of a highly resistant potato variety infected by *Phytophthora infestans*. Z: Zoospore and appressoria; N: nucleus; P.S.: protoplasmic strands; H: infected hyphae; B.M.: Brownian movement granules. (From Tomiyama, K., *Ann. Phytopathol. Soc. Jpn.*, 21, 54, 1956. With permission.)

tran-bound *p*-chloromercuribenzoic acid also inhibits the hypersensitive reaction of potato tuber cells following infection by an incompatible late blight fungus.[66] Cell browning is also inhibited by pretreatment with these reagents tested. These facts suggest that some −SH compound present in the plasma membrane of the host might be involved in recognition of the incompatible fungus.

The degradation system of rishitin, a phytoalexin of potato, begins to operate when the potato tissue is injured; but in the hypersensitively dead cells, the degradation system is inactivated and hence rishitin accumulates.[67]

4. Phytoalexins. Based on results of detailed double inoculation experiments using virulent and avirulent races of the late blight fungus and cut potato tubers, Müller and Börger[68,69] concluded that some antifungal substances might be synthesized and accumulated in the tissues which are inoculated with avirulent races, and thus presented their "phytoalexin" theory. The summary of the experiments is as follows:

1. A principle, designated as "phytoalexin", which inhibits the development of the fungus in a hypersensitive tissue, is formed or activated only when the host cells come into contact with the parasite.
2. The defensive reaction occurs only in living cells.
3. The inhibitory material is a chemical substance and may be regarded as the product of necrobiosis of the host cell.
4. This phytoalexin is nonspecific in its toxicity toward fungi; however, fungal species may be differentially sensitive to it.
5. The basic response that occurs in resistant and susceptible hosts is similar. The basis of differentiation between resistant and susceptible hosts is the speed of formation of the phytoalexin.

6. The defense reaction is confined to the tissue colonized by the fungus and its immediate neighborhood.
7. The resistant state is not inherited. It is developed after the fungus has attempted infection. The sensitivity of the host cell that determines the speed of the host reaction is specific and genotypically determined.

The same type of phenomenon was reported by Mizukami[70] in barley. Namely, detached barley leaves were injured with a glass capillary tube, and the injured parts were inoculated with a conidia suspension of a barley pathogen *(Fusarium nivale)* or a nonpathogen, *(F. solani)*. These leaves were incubated; the droplets were recovered 24 hr later and tested for the inhibitory activity for spore germination of *F. solani*. As a result, the droplets which had been inoculated with *F. solani* completely inhibited spore germination while droplets derived from the conidia suspension of *F. nivale* did not. Thus he found that the antifungal principle was induced in barley leaves by the nonpathogen of barley.

The first characterization of phytoalexin might have been on ipomeamarone. Ipomeamarone was first isolated as the bitter principle from sweet potato root infected by *Ceratosystis fimbriata,*[71] and the chemical structure of the substance was established.[72] Later on, the significance of this substance as a defense substance was confirmed.[73] Orchinol was found to be formed as a result of interaction between *Rhizoctonia repens* and tubers of *Orchis militaris.*[74] Cruickshank and Perrin[75,76] isolated phytoalexin from pea endocarps inoculated with *Monilinia fructicola* and named pisatin. The chemical structure of pisatin was established by Perrin and Bottomley.[77] From a carrot root inoculated with a nonpathogen of the carrot, *Ceratocystis fimbriata,* isocoumarin was isolated as a phytoalexin.[78] The phytoalexin which was presumed by Müller and Börger[68,69] in their system of the potato late blight disease was isolated about 30 years later by Tomiyama et al.[79] and named rishitin. The chemical structure was established by Katsui et al.[80] At the present time, more than 150 phytoalexins are chemically characterized from more than 100 plant species within about 20 families; some of them are indicated in Figure 13. The biosynthetic mechanism of phytoalexin is now of worldwide interest not only for plant pathologists, but also for plant biochemists and organic chemists.

It has been widely believed that phytoalexin may be the important factor in induced resistance as a chemical barrier, although some scientists consider that phytoalexins are only biochemical symptoms and not the cause of resistance.[81] However, the fact that introduction of the gene (transformation) for the pisatin demethylation enzyme (demethylated pisatin is less fungi toxic than pisatin) into the nonpathogenic fungus of peas renders this fungus to be pathogenic on the pea plant and shows that pisatin is an important factor for resistance.[82] This will be described later.

The biosynthesis of phytoalexins after microbial attack seems to be similar to the phenomenon of immune reaction in vertebrates. However, in contrast

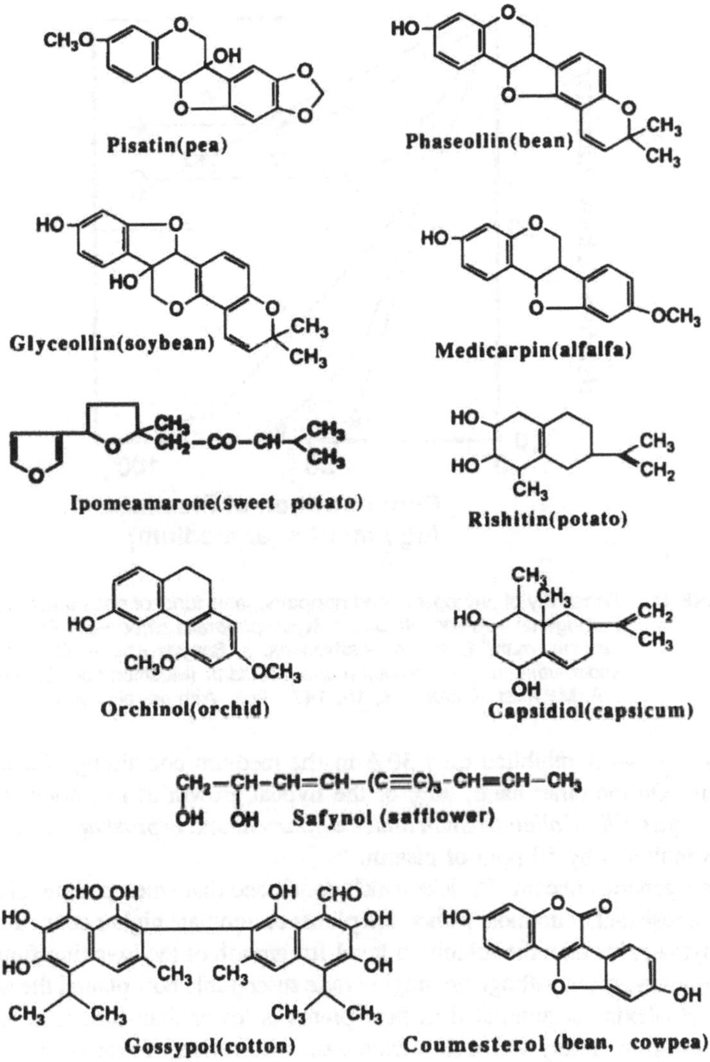

FIGURE 13. Main phytoalexins and their hosts.

to antibodies in the immune reaction in vertebrates, phytoalexins are low molecular weight compounds; and this kind of phytoalexin is host specific. Further, the antibiotic activity of phytoalexins is nonspecific to microorganisms.

Cruickshank[83] studied in detail the antifungal activity of purified pisatin and found that the pathogenic fungi of peas are more tolerant to pisatin than nonpathogenic fungi are. As indicated in Figure 14, the hyphal growth of pea pathogens — *Ascochyta pisi*, *Mycosphaerella pinodes*, and *Fusarium solani*

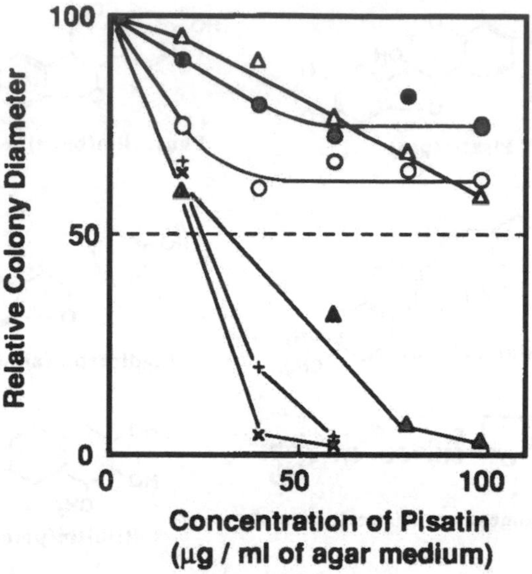

FIGURE 14. Sensitivity of pathogenic and nonpathogenic fungi of peas against pisatin. Pathogens: ● *Ascochyta pisi*, ○ *Mycosphaerella pinodes*, △ *Fusarium solani* var. *martii* f. *pisi*; nonpathogens: ▲ *Botrytis allii*, + *Colletotrichum lindemuthianum*, × *Leptosphaeria maculans*. (Modified from Cruickshank, I. A. M., *Aust. J. Biol. Sci.*, 15, 147, 1962. With permission.)

f. sp. *pisi* — is inhibited only 30% in the medium containing 100 ppm of pisatin. On the other hand, 90% of the hyphal growth of pea nonpathogens — *Botrytis allii*, *Colletotrichum lindemuthianum*, and *Leptoshaeria macurans* — is inhibited by 50 ppm of pisatin.

As a general concept, Cruickshank[84] concluded that when pathogenic fungi invade resistant or nonhost plants, the plants accumulate higher concentrations of phytoalexins than the inhibition level for growth of the invading fungi. On the contrary, when pathogenic fungi invade susceptible host plants, the amount of phytoalexins accumulated in host plants is lower than the toxic level to the pathogens. Many scientists attribute the mechanism of tolerance of pathogenic fungi against phytoalexins to the detoxifying ability of the pathogens. This will be described later.

a. Phytoalexin as the Infection Inhibitor. Oku et al.[85] found that phytoalexin inhibits the penetration of pathogenic fungus into a host plant at far below the inhibitory concentration to growth of the pathogen. That is, a pea pathogen, *Erysiphe pisi*, is highly tolerant to pisatin (ED_{50} for spore germination is 530 ppm),[86] but the treatment of pea leaves with 30 ppm of pisatin within 15 hr after inoculation inhibits markedly the infection by this fungus. Treatment later than 15 hr after inoculation does not inhibit the infection (Figure 15). The penetration of the other pea pathogen, *Mycosphaerella pi-*

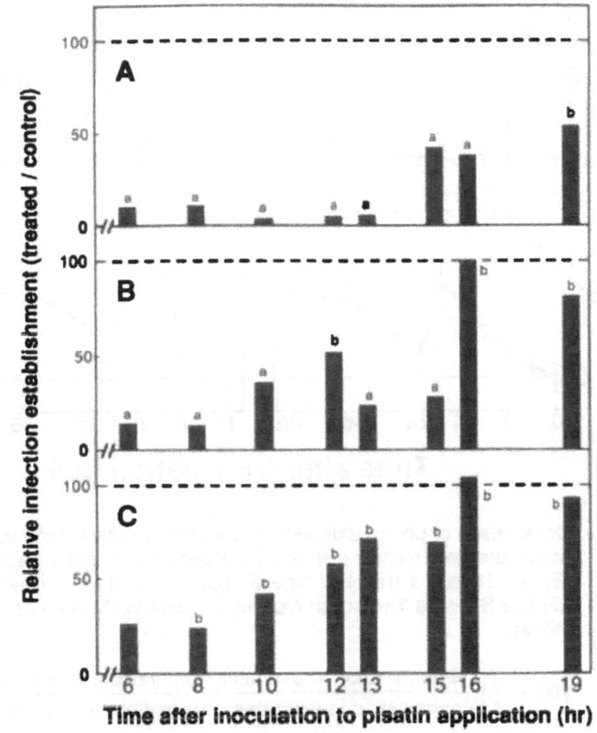

FIGURE 15. Effect of pisatin on infection establishment of *Erysiphe pisi* on pea leaves. After the lower epidermis of half leaves was stripped off, the mesophyll tissue was brought into direct contact with (A) 100 ppm, (B) 30 ppm, or (C) 10 ppm pisatin solution. Infection frequency was estimated 40 hr after pisatin administration. (a) Represents a significant difference and (b) a nonsignificant difference from control, respectively, at $p = 0.05$. (From Oku, H., Shiraishi, T., and Ouchi, S., *Ann. Phytopath. Soc. Jpn.*, 42, 599, 1976. With permission.)

nodes, is also inhibited at a lower concentration of pisatin than the growth inhibition level. These facts show that pisatin inhibits the penetration of pea powdery mildew fungus but does not inhibit the growth of the fungus after the infection has been established.

b. Two Phases of Phytoalexin Accumulation and the Role in Resistance. Chronological determination of phytoalexin activity accumulated in barley cultivars after inoculation with powdery mildew fungus, *Erysiphe graminis* f. sp. *hordei*, indicates two phases of accumulation are present as shown in Figure 16.[87] The same is true in the pea powdery mildew system.[86]

The first-phase accumulation of phytoalexin activity in barley was found at 8–20 hr after inoculation with the powdery mildew fungus on resistant cultivars, but not found in susceptible cultivars. The second-phase accumulation

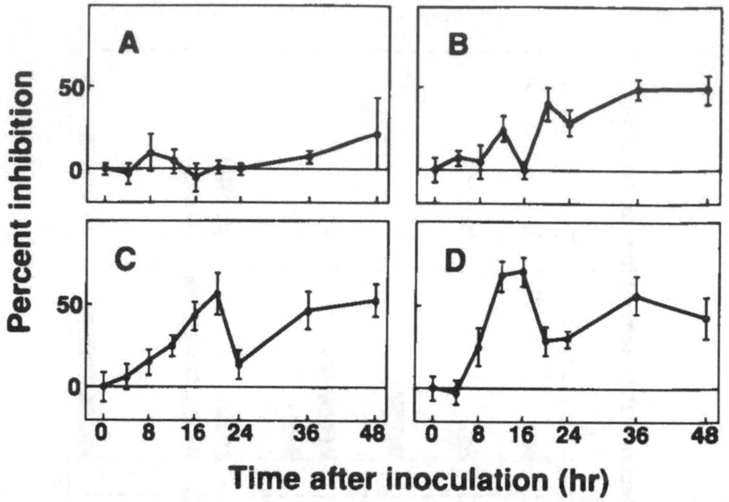

FIGURE 16. Comparison of phytoalexin induction in some cultivar-race interactions of barley powdery mildew disease. (A) Kobinkatagi-race 1 (reaction type 4), (B) no. 21-race 1 (reaction type 3), (C) no. 241-race 1 (reaction type 2), (D) H.E.S.4-race 1 (reaction type 0). Percent inhibition was calculated as follows:

$$100 - \left(\frac{\text{\% germination in exudates from inoculated leaves}}{\text{\% germination in exudates from noninoculated leaves}} \times 100\right)$$

(From Oku, H. et al., *Ann. Phytopathol. Soc. Jpn.*, 41, 187, 1975. With permission.)

was found in both susceptible and resistant cultivars after the infection had been established. Since the accumulation of the first phase just coincides with the timing of penetration, the authors consider that the first-phase accumulation plays a role in resistance to fungal penetration. The second phase one might be important in resistance to colony enlargement. The colony of powdery mildew fungus on barley leaves develops exponentially at the early stage of infection but ceases to grow at the later stage of pathogenesis. This might be due to the fact that a large amount of second-phase phytoalexin accumulates around the colony and all races of *E. graminis* tested are sensitive to the phytoalexin of barley. In contrast, the colonies of pea powdery mildew fungus continue to grow, and finally the infected leaves wither. This might be due to the tolerance of this fungus to pisatin, a phytoalexin of peas. Withering of leaves was found to be due to the toxicity of pisatin accumulated at high concentrations.[88]

Sorghum produces flavonoid phytoalexins, apigenidin and luteolinidin,[89] and caffeic ester of araninosyl-5-O-apigenidin.[90] These phytoalexins are characteristically red in color; hence their distribution within plant tissue is easily detectable. Snyder and Nicholson[91] studied the subcellular synthesis of these phytoalexins. They demonstrated that within hours after the formation of appressoria by *Colletotrichum graminicola* the underlying host cell begins to form colorless vesicle-like inclusions. Over a period of no more than 3–10 hr these inclusions move toward the site of appressorium attachment, coalesce, and become intensely pigmented. The vesicle-like inclusions finally become tightly clustered beneath the fungal appressorium and shortly thereafter burst, releasing their pigmented phytoalexins into the cell in which they were synthesized. After the inclusions burst, the water soluble phytoalexins leak out of the host cell and into the overlying appressorium causing death of the fungal germ. Thus, the phytoalexins of sorghum play a very important role at an early stage of infection.

In several bacterial diseases, the multiplication of bacteria is reported to be inhibited by phytoalexins.

Phaseollin and three other kinds of antimicrobial compounds accumulate in bean leaves infected by both compatible and incompatible halo-blight pathogens, *Pseudomonas syringae* pv. *phaseolicola* or *P. nors-purunorum.*[92] Phaseollin does not show any antibacterial activity to these pathogens at 100 μg/mL but inhibits the growth of the saprophyte, *Brevibacterium linens.* One of the other three kinds of antibiotic substances is coumesterol, but two other compounds have not been characterized. Coumesterol is not present in healthy bean leaves, but is accumulated at a concentration to inhibit bacterial multiplication 1 day after infection when the host-parasite combination is incompatible. When the disease development is severe (susceptible combinations), coumesterol accumulates 5 days after inoculation. Thus, phaseollin may not be responsible for the resistance of bean leaves against bacterial pathogens, but coumesterol and the other two unidentified compounds may play roles in inhibition of the multiplication of pathogens in hypersensitive lesions.

Coumesterol also accumulates in soybean leaves infected by the bacterial blight pathogen, *Pseudomonas syringae* pv. *glycinea,* with several other phytoalexins. Among these, glyceollin and coumesterol strongly suppress the multiplication of this pathogen and inhibit the development of the disease.[93]

The rotting of potato tubers by soft rot bacterium, *Erwinia carotovora,* has been known to be severe in low oxygen conditions; but with sufficient air supply, tubers are resistant. Lyon et al.[94,95] found that the accumulation of large amounts of phytoalexins, rishitin and phytuberin, following inoculation with *E. carotovora* occurred only when tubers were maintained in air, but not in low oxygen. Because rishitin was active against many bacteria including *E. carotovora* but phytuberin was not, they concluded that rishitin may be the main factor for resistance of potato tubers incubated in air.

5. Increase in Phenolic Components

Phenolic components, widely distributed in the plant kingdom, have been known to increase by infection with pathogens. The browning of hypersensitively

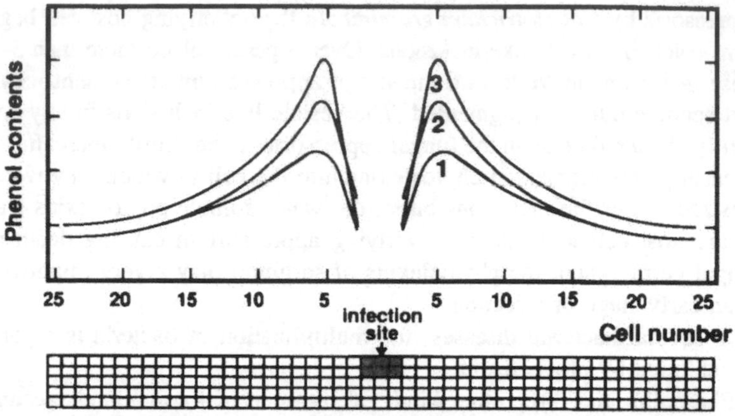

FIGURE 17. Distribution of ortho-diphenols around the infection site of an incompatible race of *Phytophthora infestans* in the potato tuber. 1, 2, and 3 represent 1, 2, and 3 days after inoculation. (Modified from Tomiyama, K., *Plant Infection Physiology*, The University of Tokyo Press, Tokyo, 1979, 57.)

reacted cells is considered to be caused by the polymerization of oxidation products of polyphenols. As described in Section II on static resistance, polyphenols have attracted the attention of many scientists, because polyphenols and oxidation products have antimicrobial activity.

Figure 17 is a schematic illustration of the distribution of *o*-diphenols near the infection site of the potato tuber inoculated with an incompatible race of the late blight fungus.[96] The infected cells accumulate low concentrations of *o*-diphenols because these cells are dead; and polyphenols are oxidized and polymerized, and hence cannot be analyzed as phenols. As seen in the figure, the content of *o*-diphenols increases very much in the living cells surrounding infected dead cells, and the growth of infected fungus may be inhibited by the increased phenols.

The same phenomenon has been known in many host-parasite combinations. In general, the increase in phenolic substances is more rapid and restricted around the infection site in resistant combinations than in susceptible combinations. The increase in the content of phenolics by infection is considered to be due to the increased biochemical activities of the hosts.

6. Postinfectionally Induced Proteins

Phenomena analogous to acquired immunity in animals have been reported in many plant virus diseases. An interferon-like substance was isolated from leaves of *Datura stramonium* infected by the tobacco mosaic virus (TMV) or tobacco necrosis virus (TNV).[97] The experimental data suggest that the substance is a protein with a molecular weight less than 30,000. At the present time, it is widely known that new proteins which are not found in healthy plants appear in many plants infected not only by viruses but also by bacterial and fungal plant pathogens.

The relationship between induced resistance and these new proteins has been the subject of extensive investigation. Many scientists have found a positive correlation between the amount of new proteins and the degree of induced resistance. These proteins are called b proteins or pathogenesis-related proteins (PR proteins). Van Loon[98] defined PR proteins as being low in molecular weight, acid soluble, low in isoelectric point, resistant to proteolytic digestion, and secreted into intercellular fluid; and having no apparent enzyme activities. However, four PR proteins were identified recently as chitinases[99] and the other four as β-1,3-glucanases.[100]

The injection of sporangiospores of *Peronospora tabacina* or inoculation with TMV has been found to induce systemic protection in tobacco against diseases caused by the same fungus and TMV.[101] The protected leaves accumulate PR proteins, and they discussed the possibility that PR proteins which have β-1,3-glucanase activity degrade the mycelial wall of *P. tabacina* distributed in the intercellular spaces of tobacco leaves. Tomato leaves infected by *Phytophthora infestans* or *Fulva fulva* also accumulate 11 PR proteins with apparent molecular weights ranging from 13 to 82 kDa.[102] Similar changes in proteins were induced by irradiation of leaves with UV light; and by injecting leaves with indole acetic acid (IAA), ethephone, fusicoccin, or an elicitor from *Phytophthora megasperma* f. sp. *glycinea*. These treated leaves increased resistance to *Phytophthora infestans*, shown by reduction in numbers of lesions and lesion size. The untreated leaves of the treated plant also acquired resistance against *P. infestans* together with the changes in protein composition. The fact that PR proteins are induced by various pathogens and chemical and physical stimuli suggests that these proteins are produced as the result of metabolic disturbance.

7. Hydroxyproline-Rich Glycoproteins

Hydroxyproline-rich glycoproteins (HRGPs) are usually found in low amounts in the cell wall of higher plants. The initial discovery of the increase in HRGPs in cell walls was in susceptible melon during infection by *Colletotrichum lagenarium*.[103,104] Since that work, it was found that HRGPs increase in the cell walls of other hosts infected by various pathogens; and the accumulation is, in general, earlier after inoculation in resistant plants than in susceptible ones.[105,106] Thus, HRGPs might be involved in defense reactions in many plants.[105,107,108]

One third of HRGPs in cell walls are composed of protein, and the remaining two thirds are composed of polysaccharide. About 40% of proteins is composed of hydroxyproline, and the polysaccharide moiety is composed mainly of arabinose. Studies on an isolated HRGP have confirmed that its molecule is linear[109] and is rich in basic amino acids.[108] These features confer two properties. First, because it is a linear molecule, it may function as a structural polymer and strengthen the cell wall. Second, the high level of basic amino acids confers the properties of a polycation, and it may agglutinate negatively charged particles or cells of pathogens. Both properties provide the basis for a role in a resistance mechanism.[106] The increase of HRGPs is more prominent in incompatible host-parasite combinations than in compatible ones. No significant increase is found in infected monocotyledons.[106]

8. Unsaturated Fatty Acids and the Oxidation Products

Unsaturated fatty acids released from the plasma membrane by some stimuli have antibiotic activity. The oxidation of the fatty acids further increases the antibiotic activity. Epoxy fatty acids were isolated from resistant cultivars of the rice plant against the blast fungus as defense substances.[110]

Yamamoto and Tani[111] suggested the involvement of lypoxygenase in the resistance of oat cultivars against *Puccinia coronata*. That is, oat leaves inoculated with an incompatible race of the fungus produce specific lipoxygenase isozymes, and the oxidation of linoleic acid by the lypoxygenase isozyme gave similar Rf values on the thin-layer chromatographic (TLC) plate as an antibiotic substance which was extracted from oat leaves inoculated with an incompatible race of the fungus. The appearance of the specific isozyme was inhibited by treatment of leaves with inhibitors of nucleic acid synthesis or protein synthesis.

9. Superoxide Radical

The inoculation of potato tissue with an incompatible race of ***Phytophthora infestans*** induces the superoxide (O_2^-) generation.[112,113] The superoxide itself does not show antibiotic activity but is responsible for triggering hypersensitive reaction and phytoalexin biosynthesis. The amount of superoxide produced by potato tissue is larger in incompatible potato-late blight combinations than in compatible ones.

B. Systemic Induced Resistance

In many plant-pathogen combinations, it is well-known that infection of part of a plant by viral, bacterial, and fungal pathogens induces resistance to the other parts of the same plant against a variety of pathogens. This type of phenomenon is called systemic induced resistance. Although the mechanism of systemic induced resistance is not fully understood, this phenomenon has been used practically as a very effective control measure for some virus diseases such as citrus tristeza disease, tomato mosaic disease, and so on.

Further, according to recent public opinion against pesticide pollution, many scientists are conducting research to apply the phenomena and/or mechanisms of systemic induced resistance to develop effective control measures of plant diseases; and the clarification of the mechanism seems to be very promising to the control of plant diseases. Therefore, this will be described in detail in Chapter 3 in relation to new control measures for plant diseases.

REFERENCES

1. Melander, L. W. and Craigie, J. H., Nature of resistance of *Berberis* spp. to *Puccinia graminis*, *Phytopathology*, 17, 95, 1927.
2. Sakamoto, M., Studies on the resistance of rice plant against blast disease, in *Jubilee Publication in Commemoration of 60th Birthday of Professor M. Sakamoto*, Publication Committee, Tohoku University, Sendai, 1968, 1.
3. Yoshii, H., Pathological studies on rice blast caused by *Pyricularia oryzae*. II. On the mode of infection of the pathogen, *Ann. Phytopathol. Soc. Jpn.*, 6, 205, 1936.
4. Akai, S., Histology of defense in plants, in *Plant Pathology I*, Horsfall, J. G. and Dimond, A. E., Eds., Academic Press, New York, 1959, 391.
5. Hart, T., Relation of stomatal behaviour to stem rust resistance in wheat, *J. Agric. Res.*, 39, 929, 1929.
6. Pool, V. W. and McKay, M. B., Relation of stomatal movement to infection by *Cercospora beticola*, *J. Agric. Res.*, 22, 1011, 1916.
7. Caldwell, R. M. and Stone, G. M., Relation of stomatal function of wheat to invasion and infection by leaf rust *(Puccinia triticina)*, *J. Agric. Res.*, 52, 917, 1936.
8. Hirt, R. R., Relation of stomata to infection of *Pinus strobus* by *Cronartium libicola*, *Phytopathology*, 28, 180, 1938.
9. Akai, S., *Helminthosporium* blight of rice plant, the special reference to pathological physiology of the affected plants, *Ann. Phytopathol. Soc. Jpn.*, 31, 193, 1965.
10. Kosaka, T., Ecology of *Pellicularia* sheath-blight of rice plant and its chemical control, *Ann. Phytopathol. Soc. Jpn.*, 31, 179, 1965.
11. Ono, K., Ecology of stem rot of rice and its control, *Ann. Phytopathol. Soc. Jpn.*, 31, 173, 1965.
12. Tokunaga, Y., Ecology of rice blast, *Ann. Phytopathol. Soc. Jpn.*, 31, 165, 1965.
13. Ingham, J., Concept of pre-infectional and post-infectional resistance, *Phytopathol. Z.*, 78, 314, 1973.
14. Link, K. P., Angell, H. R., and Walker, J. C., The isolation of protocatechuic acid from pigmented onion scales and its significance in relation to disease resistance in onions, *J. Biol. Chem.*, 81, 369, 1929.
15. Link, K. P., Dickson, A. D., and Walker, J. C., Further observation on the occurrence of protocatechuic acid in pigmented onion scales and its relation to disease resistance in the onion, *J. Biol. Chem.*, 84, 719, 1929.
16. Link, K. P. and Walker, J. C., The isolation of catechol from pigmented onion scales and its significance in relation to disease resistance in onions, *J. Biol. Chem.*, 100, 379, 1933.
17. Johnson, G. and Schaal, L. A., Relation of chlorogenic acid to scab resistance in potatoes, *Science*, 115, 627, 1952.

18. Oku, H., Biochemical studies on *Cochliobolus miyabeanus*. IV. Fungicidal action of polyphenols and the role of polyphenoloxidase of the fungus, *Phytopathol. Z.*, 38, 342, 1960.
19. Oku, H., Biochemical studies on *Cochliobolus miyabeanus*. III. Some oxidizing enzymes of the rice plant and its parasites and their contribution to the formation of the lesions, *Ann. Phytopathol. Soc. Jpn.*, 23, 169, 1958.
20. Lyl, H., On the toxicity of oxidized phenols, *Phytopathol. Z.*, 52, 229, 1965.
21. Alberghina, F. A. M., On enzyme inhibition by oxidized chlorogenic acid, *Life Sci.*, 3, 49, 1964.
22. Patil, S. S. and Dimond, A. E., Inhibition of *Verticillium* polygalacturonase by oxidation products of polyphenols, *Phytopathology*, 57, 429, 1967.
23. Kosuge, T., The role of phenolics in host response to infection, *Annu. Rev. Phytopathol.*, 7, 195, 1969.
24. Schonbeck, F. and Schlosser, E., Preformed substances as potential protectant, in *Physiological Plant Pathology*, Heitefuss, R. and Williams, P. H., Eds., Springer-Verlag, Berlin, 1976, 653.
25. Williams, E. B. and Kuć, J., Resistance in *Malus* to *Venturia inaquales, Annu. Rev. Phytopathol.*, 7, 223, 1969.
26. Durbin, R. D. and Uchytil, T. F., The role of allicin in the resistance of garlic to *Penicillium* spp., *Phytopathol. Mediterr.*, 10, 227, 1971.
27. Durbin, R. D. and Uchytil, T. F., Purification and properties of alliin lyase from the fungus *Penicillium corymbiferum, Biochim. Biophys. Acta*, 229, 518, 1971.
28. Bagham, A. D., Horne, R. W., Glauert, A. M., Dingle, J. T., and Lucy, J. A., Action of saponin on biological cell membrane, *Nature (London)*, 196, 952, 1962.
29. Arneson, P. A. and Durbin, R. D., The sensitivity of fungi to tomatine, *Phytopathology*, 58, 536, 1968.
30. Arneson, P. A. and Durbin, R. D., Studies on the mode of action of tomatine as a fungitoxic agent, *Plant Physiol.*, 43, 683, 1968.
31. Allen, P. J. and Kuć, J., Solanine and chaconine as fungitoxic compounds in extracts of Irish potato tubers, *Phytopathology*, 58, 776, 1968.
32. Turner, E. M. C., An enzymic basis for pathogenic specificity in *Ophiobolus graminis*, *J. Exp. Bot.*, 12, 169, 1961.
33. Stoessl, A., The antifungal factors in barley. IV. Isolation, structure and synthesis of the hordatins, *Can. J. Chem.*, 45, 1745, 1967.
34. Stoessl, A. and Unwin, C. H., The antifungal factors in barley. V. Antifungal activity of the hordatins, *Can. J. Bot.*, 48, 465, 1970.
35. Huang, P.-Y., Huang, J. S., and Goodman, R. N., Resistance mechanisms of apple shoots to an avirulent strains of *Erwinia amylovora, Physiol. Plant Pathol.*, 6, 283, 1975.
36. Goodman, R. N., Huang, R., and White, J. A., Ultrastructural evidence of immobilization of an incompatible bacterium, *Pseudomonas pisi*, in tobacco leaf tissue, *Phytopathology*, 66, 754, 1976.
37. Huang, J. S. and Van Dyke, C. G., Interaction of tobacco callus tissue with *Pseudomonas tabaci, P. pisi* and *P. fluorescens, Physiol. Plant Pathol.*, 13, 65, 1978.
38. Romeiro, R., Karr, A., and Goodman, R. N., Isolation of a factor from apple that agglutinate *Erwinia amylovora, Plant Physiol.*, 68, 772, 1981.

39. Romeiro, R., Karr, A., and Goodman, R. N., *Erwinia amylovora* cell wall receptor for apple agglutinin, *Physiol. Plant Pathol.*, 19, 383, 1981.
40. Sequeira, L. and Graham, T. L., Agglutination of avirulent strains of *Pseudomonas solanacearum* by potato lectin, *Physiol. Plant Pathol.*, 11, 43, 1977.
41. Uritani, I. and Kojima, M., Spore agglutinating factor and germ tube growth inhibiting factor in host plant, in *Recognition and Specificity in Plant Host-Parasite Interactions*, Daly, J. M. and Uritani, I., Eds., Japan Scientific Society Press, Tokyo/University Park Press, Baltimore, 1979, 181.
42. Hohl, H. R., Balsiger, S., Odermatt, M., Werner, C., and Guggenbuhl, C., Interactions of *Phytophthora* with protoplasts of its host, *Abstr. Pap. 5th ICPP*, Kyoto, 1988, 218.
43. Kano, H., Haga, M., Aoyagi, T., and Sekizawa, Y., Possible participation of rice leaf lectin in the reception of blast fungus proteoglucomannan onto the host cells, *Abstr. Pap. 5th ICPP*, Kyoto, 1988, 251.
44. Garas, N. A. and Kuć, J., Potato lectin lyses zoospores of *Phytophthora infestans* and precipitates elicitors of terpenoid accumulation produced by the fungus, *Physiol. Plant Pathol.*, 18, 227, 1981.
45. Andreu, A. and Daleo, R. G., Properties of potato lectin fractions isolated from different parts of the tuber and their effect on the growth of *Phytophthora infestans*, *Physiol. Mol. Plant Pathol.*, 32, 323, 1988.
46. Doke, N., Garas, N., and Kuć, J., Effect on host hypersensitivity of suppressors released during the germination of *Phytophthora infestans* cystospores, *Phytopathology*, 70, 35, 1980.
47. De Bary, A., Recherches sur le developpement de quelques champignons parasites, *Ann. Sci. Nat. Bot. Biol. Veg.*, 20, 5, 1863.
48. Aist, J. R., Papillae and related wound plugs of plant cells, *Annu. Rev. Phytopathol.*, 14, 145, 1976.
49. Sherwood, R. and Vance, C., Initial events in the epidermal layer during penetration, in *Plant Infection: The Physiological and Biochemical Basis*, Asada, Y., Bushnell, W., Ouchi, S., and Vance, C., Eds., Japan Scientific Society Press, Tokyo, 1982, 27.
50. Stumm, D. and Gessler, C., Role of papillae in the induced systemic resistance of cucumbers against *Colletotrichum lagenarium*, *Physiol. Plant Pathol.*, 29, 405, 1986.
51. Kita, N., Toyoda, H., and Shishiyama, J., Chronological responses in powdery-mildewed barley leaves, *Can. J. Bot.*, 59, 1761, 1981.
52. Mayama, S. and Shishiyama, J., Localized accumulation of fluorescent and U.V.-absorbing compounds at penetration sites in barley infected with *Erysiphe graminis hordei*, *Physiol. Plant Pathol.*, 13, 347, 1978.
53. Asada, Y. and Matsumoto, I., The nature of lignin obtained from downy mildew-infected Japanese radish root, *Phytopathol. Z.*, 73, 208, 1972.
54. Asada, Y., Oguchi, T., and Matsumoto, I., Induction of lignification in response to fungal infection, in *Recognition and Specificity in Plant Host-Parasite Interactions*, Daly, J. M. and Uritani, I., Eds., Japan Scientific Society Press, Tokyo/University Park Press, Baltimore, 1979, 99.
55. Hammerschmidt, R. and Kuć, J., Lignification as a mechanism for induced systemic resistance in cucumber, *Physiol. Plant Pathol.*, 20, 61, 1982.
56. Dean, R. A. and Kuć, J., Rapid lignification in response to wounding and infection as a mechanism for induced systemic protection in cucumber, *Physiol. Plant Pathol.*, 31, 69, 1987.

57. Grand, C. and Rossignol, M., Changes in the lignification process induced by localized infection of muskmelons by *Colletotrichum lagenarium*, *Plant Sci. Lett.*, 28, 103, 1982.
58. Southernton, S. G. and Deverall, B. J., Histochemical and chemical evidence for lignin accumulation during expression of resistance to leaf rust fungi in wheat, *Physiol. Plant Pathol.*, 36, 483, 1990.
59. Vance, C. P., Kirk, T. K., and Sherwood, R. T., Lignification as a mechanism of disease resistance, *Annu. Rev. Phytopathol.*, 18, 259, 1980.
60. Ward, H. M., On the relations between host and parasite in the bromes and their brown rust, *Puccinia dispersa* (Erikss.), *Ann. Bot.*, 16, 233, 1902.
61. Tomiyama, K., Cell physiological studies on the resistance of potato plants to *Phytophthora infestans*. IV. On the movements of cytoplasm of the host cell induced by the invasion of *Phytophthora infestans*, *Ann. Phytopathol. Soc. Jpn.*, 21, 54, 1956.
62. Tomiyama, K., Cell physiological studies on the resistance of potato plants to *Phytophthora infestans*. V. Effect of 2,4-dinitrophenol upon the hypersensitive reaction of potato plant cell to infection by *P. infestans*, *Ann. Phytopathol. Soc. Jpn.*, 22, 75, 1957.
63. Tomiyama, K. and Sakai, R., Hypersensitive response of infected cells with special reference to potato-late-blight, *Ann. Phytopathol. Soc. Jpn.*, 31, 341, 1965.
64. Nozue, M., Tomiyama, K., and Doke, N., Effect of adenosine-5′-triphosphate on hypersensitive death of potato tuber cells infected by *Phytophthora infestans*, *Phytopathology*, 68, 873, 1978.
65. Tomiyama, K., Sato, K., and Doke, N., Effect of cytochalasin B and colchicine on hypersensitive death of potato cells infected by incompatible race of *Phytophthora infestans*, *Ann. Phytopathol. Soc. Jpn.*, 48, 228, 1982.
66. Doke, N. and Tomiyama, K., Effect of sulfhydryl-binding compounds on hypersensitive death of potato tuber cells following infection with an incompatible race of *Phytophthora infestans*, *Physiol. Plant Pathol.*, 12, 133, 1978.
67. Sakai, S., Doke, N., and Tomiyama, K., Relation between necrosis and rishitin accumulation in potato tuber slices treated with hyphal wall components of *Phytophthora infestans*, *Ann. Phytopathol. Soc. Jpn.*, 48, 238, 1982.
68. Müller, K. O. and Börger, H., Studien uber den "Mechanismus" der *Phytophthora*-Resistenz der Kartoffel, *Landwirtsch. Jahrb. Berlin*, 87, 609, 1939.
69. Müller, K. O. and Börger, H., Experimentelle Untersuchungen uber die *Phytophthora*-Resistenz der Kartoffel, *Arb. Biol. Reichsanst. Land Forstwertsch. Berlin*, 23, 189, 1940.
70. Mizukami, T., Observation on the reactions of plant to the infection of some pathogens. I. On the difference of the influence of the barley juice on the conidial germination of *Fusarium nivarle* and *Fusarium solani*, *Ann. Phytopathol., Soc. Jpn.*, 17, 57, 1953.
71. Hiura, M., Studies in storage and rot of sweet potato, *Gifu Norin Semmon Gakko Gakujutsu Hokoku*, 50, 1, 1943.
72. Kubota, T. and Matsuura, T., Chemical studies on the black spot disease of sweet potato, *J. Chem. Soc. Jpn., Pure Chem. Sect.*, 74, 248, 1953.
73. Uritani, I. and Akazawa, T., Antibiotic effect on *Ceratostomella fimbriata* of ipomeamarone, an abnormal metabolite in black rot of sweet potato, *Science*, 121, 216, 1955.

74. Gäumann, E., Broun, R., and Bazzigher, G., Uber indizierte Abwehrreaktionen by Orchideen, *Phytopathol. Z.*, 17, 36, 1950.
75. Cruickshank, I. A. M. and Perrin, D. R., Isolation of phytoalexin from *Pisum sativum* L., *Nature (London)*, 187, 799, 1960.
76. Cruickshank, I. A. M. and Perrin, D. R., Studies on phytoalexins. III. The isolation, assay and general properties of phytoalexin from *Pisum sativum* L., *Aust. J. Biol. Sci.*, 14, 336, 1961.
77. Perrin, D. R. and Bottomley, W., Pisatin: an antifungal substance from *Pisum sativum* L., *Nature (London)*, 191, 76, 1961.
78. Condon, P. and Kuć, J., Isolation of a fungitoxic compound from carrot root tissue inoculated with *Ceratocystis fimbriata*, *Phytopathology*, 50, 267, 1960.
79. Tomiyama, K., Sakai, R., Ishizuka, N., Sato, N., Katsui, N., Takasugi, M., and Masamune, T., A new antifungal substance isolated from resistant potato tuber tissue infected by pathogens, *Phytopathology*, 58, 115, 1968.
80. Katsui, N., Murai, A., Takasugi, M., Imaizumi, K., Masamune, T., and Tomiyama, K., The structure of rishitin, a new antifungal compound from diseased potato tubers, *J. Chem. Soc. D*, 1968, 43, 1968.
81. Király, Z., Érsek, T., and Adam, A., Phytoalexins: the question of their role in disease resistance, *Abstr. Pap. 5th ICPP*, Kyoto, 1988, 223.
82. VanEtten, H. D., Mathews, D. E., and Mathews, P. S., Phytoalexin detoxification: importance for pathogenicity and practical implications, *Annu. Rev. Phytopathol.*, 27, 143, 1989.
83. Cruickshank, I. A. M., Studies on phytoalexins. IV. The antimicrobial spectrum of pisatin, *Aust. J. Biol. Sci.*, 15, 147, 1962.
84. Cruickshank, I. A. M., Phytoalexins, *Annu. Rev. Phytopathol.*, 1, 351, 1963.
85. Oku, H., Shiraishi, T., and Ouchi, S., Effect of preliminary administration of pisatin to pea leaf tissues on the subsequent infection by *Erysiphe pisi* DC, *Ann. Phytopathol. Soc. Jpn.*, 42, 597, 1976.
86. Oku, H., Ouchi, S., Shiraishi, T., and Baba, T., Pisatin production in powdery mildewed pea seedlings, *Phytopathology*, 65, 1263, 1975.
87. Oku, H., Ouchi, S., Shiraishi, T., Komoto, Y., and Oki, K., Phytoalexin activity in barley powdery mildew, *Ann. Phytopathol. Soc. Jpn.*, 41, 185, 1975.
88. Shiraishi, T., Oku, H., Isono, M., and Ouchi, S., The injurious effect of pisatin on plasma membrane of pea, *Plant Cell Physiol.*, 16, 939, 1975.
89. Nicholson, R. L., Kollipara, S. S., Vincent, J. R., Lyons, P. C., and Cadena-Gomez, G., Phytoalexin synthesis by sorghum coleoptil in response to infection by pathogenic and nonpathogenic fungi, *Proc. Natl. Acad. Sci. U.S.A.*, 84, 5520, 1987.
90. Hipskind, J., Hanau, R., and Nicholson, R. L., Phytoalexin synthesis in sorghum: identification of an apigenin acyl ester, *Physiol. Plant Pathol.*, 36, 381, 1990.
91. Snyder, B. A. and Nicholson, R. L., Synthesis of phytoalexins in sorghum as a site specific response to fungal ingress, *Science*, 248, 1637, 1990.
92. Lyon, F. M. and Wood, R. K. S., Production of phaseollin, coumesterol and related compounds in bean leaves inoculated with *Pseudomonas* spp., *Physiol. Plant Pathol.*, 6, 117, 1975.
93. Keen, N. T. and Kennedy, B. W., Hydroxyphaseollin and related isoflavonoids in the hypersensitive resistance reaction in soybeans to *Pseudomonas glycinea*, *Physiol. Plant Pathol.*, 4, 173, 1974.

94. Lyon, G. D., Lund, B. M., Bayliss, C. E., and Wyatt, G. M., Resistance of potato tubers to *Erwinia carotovora* and formation of rishitin and phytuberin in infected tissue, *Physiol. Plant Pathol.*, 6, 43, 1975.
95. Lyon, G. D. and Bayliss, C. E., The effect of rishitin on *Erwinia carotovora* var. *atroseptica* and other bacteria, *Physiol. Plant Pathol.*, 6, 177, 1975.
96. Tomiyama, K., Change of phenol metabolism, in *Plant Infection Physiology*, The University of Tokyo Press, Tokyo, 1979, 57.
97. Loebenstein, G. and Van Praagh, T., Extraction of a virus interfering agent induced by localized and systemic infection, in *Host-Parasite Relations in Plant Pathology*, Király, Z. and Ubrizsy, G., Eds., Research Institute Plant Protection, Budapest, 1964, 53.
98. Van Loon, L. C., Pathogenesis-related proteins, *Plant Mol. Biol.*, 4, 111, 1985.
99. Legrand, M., Kaufman, S., Geoffroy, P., and Fritig, B., Biological function of pathogenesis-related proteins: four tobacco pathogenesis-related proteins are chitinases, *Proc. Natl. Acad. Sci. U.S.A.*, 34, 6750.
100. Kaufman, S., Legrand, M., Geoffroy, P., and Fritig, B., Biological function of pathogenesis-related proteins: four PR-proteins of tobacco have 1,3-glucanase activity, *EMBO J.*, 6, 3206.
101. Ye, X. S., Pan, S. Q., and Kuć, J., Pathogenesis-related proteins and systemic resistance to blue mould and tobacco mosaic virus induced by tobacco mosaic virus, *Peronospora tabacina* and aspirin, *Physiol. Plant Pathol.*, 35, 161, 1989.
102. Christ, U. and Mosinger, F., Pathogenesis-related proteins of tomato. I. Induction by *Phytophthora infestans* and other biotic and abiotic inducers and correlation with resistance, *Physiol. Mol. Plant Pathol.*, 35, 53, 1989.
103. Esquérré-Tugayé, M. T., Influence d'une maladie parasitaire sur la Leneuer en hydroxyproline des parois cellulaires d'epicotyles et petioles de plantes de Melon, *C. R. Acad. Sci. Paris, Ser. D*, 273, 525, 1973.
104. Esquérré-Tugayé, M. T. and Mazeau, D., Effect of a fungal disease on extensin, the plant cell wall glycoprotein, *J. Exp. Bot.*, 25, 50, 1984.
105. Esquérré-Tugayé, M. T., Laffitte, C., Mazeau, D., Toppan, A., and Touze, A., Cell surfaces in plant-microorganism interaction. II. Evidence for the accumulation of hydroxyproline-rich glycoproteins in the cell wall of diseased plants as a defense reaction, *Plant Physiol.*, 64, 320, 1979.
106. Mazeau, K. and Esquérré-Tugayé, M. T., Hydroxyproline-rich glycoprotein accumulation in the cell walls of plants infected by various pathogens, *Physiol. Plant Pathol.*, 29, 147, 1986.
107. Mellon, J. E. and Helgeson, J. P., Interaction of hydroxyproline-rich glycoprotein from tobacco callus with potential pathogens, *Plant Physiol.*, 70, 401, 1982.
108. Leach, J. E., Cantrell, M. A., and Sequeira, L., Hydroxyproline-rich bacterial agglutinine from potato, *Plant Physiol.*, 70, 1353, 1982.
109. Van Holst, G. J. and Varner, J. E., Reinforced polyproline. II. Conformation in a hydroxyproline-rich cell wall glycoprotein from carrot root, *Plant Physiol.*, 74, 247, 1984.
110. Kato, T., Yamaguchi, Y., Uehara, T., Yokoyama, T., Namai, T., and Yamanaka, S., Self defensive substances in rice plant against rice blast disease, *Tetrahedron Lett.*, 24, 2715, 1983.

111. Yamamoto, H. and Tani, T., Possible involvement of lypoxigenase in the mechanism of resistance of oats to *Puccinia coronata avenae*, *J. Phytopathol.*, 116, 329, 1986.
112. Doke, N., Involvement of superoxide anion generation in the hypersensitive response of potato tuber tissues to infection with an incompatible race of *Phytophthora infestans* and to the hyphal wall components, *Physiol. Plant Pathol.*, 23, 345, 1983.
113. Doke, N., Chai, H. B., and Kawaguchi, A., Biochemical basis of triggering and suppression of hypersensitive cell response, in *Molecular Determinants of Plant Disease*, Nishimura, S., Vance, C. A., and Doke, N., Eds., Japan Scientific Society Press, Tokyo/Springer-Verlag, Berlin, 1987, 235.

CHAPTER 3

Defense and Offense Between Higher Plants and Microbes and the Mechanism

The intrinsic importance in resistance of plants against pathogens is active defense as described in Chapter 2.

The first requisite for plant pathogenic microorganisms is the ability to enter into plants through protective barriers, because they have to derive food materials from their host plants. The second step of the most important task after entering into plants is to overcome resistant mechanisms of host plants, especially active defense expressed after penetration.

Hyphae of the rice blast fungus inoculated on the nonhost plant, tomato leaves, are killed immediately after penetration through the epidermis as described in Chapter 1. Plant pathogenic fungi, in general, grow freely and show no specificity of many culture media. This is due to the fact that artificial media are not living beings and do not have defense mechanisms. However, they can parasitize and grow only on their own living host or hosts. This phenomenon clearly shows that pathogens can overcome defense reactions expressed by their living hosts, but not by nonhost plants.

Because active defense is induced after the attack by microorganisms, there must be some mechanisms in plants that recognize some molecules produced during the infection process of the microbes. Conversely, how do pathogenic microorganisms overcome the defense reaction of their own host? The elucidation of these mechanisms at the molecular level might be significant not only from the scientific point of view, but also for the development of new methods of plant disease control.

In this chapter, the mechanism of elicitation of defense reactions in higher plants and then the mechanisms of pathogens to overcome the defense reactions expressed by host plants will be discussed.

I. MECHANISM OF ELICITATION OF DEFENSE REACTION IN PLANTS

A. Elicitor, Triggering Substance for Defense Reaction

In the early stage of research on phytoalexins, the accumulation of these antibiotics was reported to be induced by cell-free spore germination fluid of pathogenic fungi,[1] metal ions,[1] low molecular weight microbial metabolites such as ascochitine and ophiobolin,[2] some fungicides,[2,3] some metabolic inhibitors,[4] and ultraviolet irradiation.[5] Cruickshank and Perrin[6] first isolated and partially characterized a high molecular weight peptide which induces phaseollin biosynthesis in the kidney bean from mycelia of *Monilinia fructicola* and named monilicolin A. Later, a variety of polysaccharide components, glycoproteins, etc. that induce phytoalexin biosynthesis has been isolated from cell wall fractions, mycelia, and culture filtrates of pathogenic fungi.

These substances which induce the accumulation of phytoalexins were first termed "elicitors" by Keen.[7] However, at present the term elicitor is used more widely, that is, for substances which induce defense reactions including phytoalexin, hypersensitive reactions, and so on in higher plants. Phytoalexins and their elicitors were reviewed by Dervill and Albersheim.[8]

Elicitors are sometimes classified as biotic and abiotic elicitors. Biotic elicitors mean those of biological origin such as fungal products (exogenous elicitors) or substances released by plants as the result of host-parasite interactions (endogenous elicitors). Abiotic elicitors are of abiotic origin such as UV-light, metal ions, artificially synthesized substances, and so on. However, this classification includes some contradictions. For example, many organic substances of biological origin can be chemically synthesized. In such cases the fungal metabolite such as ascochitine — a toxic metabolite extracted from a culture filtrate of *Ascochyta fabae*[2] — is a biotic elicitor, but the chemically synthesized ascochitine is an abiotic elicitor.

Nevertheless, the so-called biotic elicitors are important in elucidating the mechanism of the defense reaction, and many exogenous elicitors have been isolated from a variety of microorganisms. They are chemically classified into polypeptides, polysaccharides, glycoproteins, protein-lipid-sugar complexes, chitosan, unsaturated fatty acids such as eicosapentanoic acid arachidonic acid, and so on.

As an example of an endogneous elicitor, Asada[9] found that the lignification-inducing factor — a glycopeptide — is bound to the cell walls of healthy radishes in an inactive form, but is liberated as an active form from cell walls by the fungal infection or other stimuli and induces the formation of diseased lignin in the infected tissue. By using their system of soybean-*Phytophthora megasperma* f. sp. *glycinea*, Keen and Yoshikawa[10] and Yoshikawa et al.[11] demonstrated that the elicitor molecules are released from cell walls of the pathogenic fungus by β-1,3-endoglucanase present in soybeans. ^{13}C-nuclear magnetic resonance (NMR) and sugar analysis indicated the presence of β-1,6- and β-1,3-glucose linkages for various released elicitors with different molecular weight.[12] Since the treatment of released elicitors with β-1,3-endoglucanase no longer causes the loss of elicitor activity, the released elicitors are mainly composed of β-1,6-glucose linkage. However, β-1,3-exoglucanase completely inactivates the elicitor activity. From these results, they concluded that the released elicitors are composed mainly of a 1,6-linkage with one to two β-1,3-glucose linkages which bound originally to the fungal cell wall.

From the available information, elicitors can be classified into nonspecific and specific elicitors. Nonspecific elicitors induce defense reactions nonspecifically to all varieties or cultivars of a host plant species, and sometimes to nonhost plant species. According to Ayres et al.,[13] glucan elicitors obtained from three races of *Phytophthora megasperma* f. sp. *glycinea* have similar structures and no specificity can be found in the activity to induce glyceollin when tested with a variety of host cultivars. Cline et al.[14] reported that glucan elicitors isolated from a soybean pathogen, *P. megasperma* f. sp. *glycinea*, induce phytoalexins not only in all host cultivars tested but also in nonhost plants such as potato and bean plants. The accumulation of pisatin in pea leaves can be induced by a polysaccharide elicitor isolated from spore germination fluid of a pea pathogen, *Mycosphaerella pinodes*, and also elicitors from nonpathogens, *M. melonis* and *M. ligulicola*.[15] Bean leaves also react with glucans extracted both from its pathogenic fungus, *Colletotrichum lindemuthianum*, and nonpathogenic fungus, *Fusarium oxysporum*.[16]

Specific elicitors produced by some races of pathogens induce defense reactions to incompatible varieties or cultivars of the host species, but do not induce or weakly induce defense reactions to the compatible ones.

Races of *Colletotrichum lindemuthianum* produce several extracellular components having elicitor activity.[17,18] The predominant component of a race elicitor has a molecular weight of about 60 kDa, and is composed of mannose (45%), galactose (17%), and glucose (38%).[18] This component displays a high level of elicitor activity on a race incompatible cultivar, Dark Red kidney bean, but has no elicitor activity on a compatible cultivar, Great Northern bean. Low levels of elicitor activity were detected on both of the above cultivars by components from compatible races. Cell wall glucans and some of the purified extracellular components from a race of *C. lindemuthianum* are recognized as elicitors by compatible Great Northern cultivars as well as incompatible Dark Red kidney beans.[18] That is, even in a compatible interaction, there are components of the fungus that trigger a defense response.

The glucan elicitor which is released from the cell wall of *Phytophthora megasperma* f. sp. *glycinea* by the activity of β-1,3-glucanase present in soybean tissue is also cultivar specific; that is, it produces a high level of glyceollin in incompatible cultivars of soybeans.[11]

Glycoprotein elicitors extracted from the cell wall of many races of *Phytophthora megasperma* f. sp. *glycinea* were revealed to be cultivar specific. The elicitor activity of these glycoproteins is believed to be due to protein moiety.[19]

Daniel et al.[20] proved in their system of chickpeas and *Ascochyta rabiei*, that resistance or susceptibility of cultivars against the pathogen is determined by the response of cultivars to the fungal elicitor. That is, they conducted experiments using cell-suspension cultures of chickpeas derived from resistant and susceptible cultivars and an elicitor from the fungus; and found that the amount of phytoalexin accumulation is determined by the degree of activation of some rate-limiting enzymes involved in phytoalexin biosynthesis, though the activation of the other enzymes is similar. The resistant and susceptible cultivars of the chickpea respond differently against the yeast glucan elicitor.[21]

In cases where elicitors produced by the pathogenic fungi have specificity to incompatible cultivars but not to compatible cultivars, the race-cultivar specificity is explained by the activity of specific elicitors as hypothesized by Keen.[22] However, the other determinant of host-parasite specificity might be involved in diseases where nonspecific elicitors are produced by pathogens at the host-parasite interfaces. These examples will be discussed later.

In most experiments on exogenous elicitors, compounds having elicitor activity have been isolated by hard treatment such as autoclaving of the cell wall fraction of pathogenic fungi or extraction with an ether-chloroform mixture, and so on. However, there is no guarantee that these substances actually act as elicitors at the host-parasite interfaces. Shiraishi et al.[23] obtained an elicitor from the spore germination fluid of *Mycosphaerella pinodes*, which induces pisatin biosynthesis in pea leaves. The substance is a polysaccharide with a relative molecular weight of ca. 70,000 Da.

B. Mechanism of Recognition of Elicitors by Host Cells

As to the recognition of elicitors by host cells, two hypothetical models have been proposed. One is that elicitors directly bind to deoxyribonucleic acid (DNA) in plant nuclei and activate specifically the transcription of genes necessary for phytoalexin biosynthesis.[24] This model is dependent on the facts that DNA-binding compounds such as actinomycin D[25,26] or UV-irradiation which forms a pyridine dimer[27] induce the phytoalexins. However, direct evidence that fungal elicitors diffuse into the plant cell nuclei and bind to DNA has not been obtained, except in the case of chitosan released from the cell wall of *Fusarium* fungus.[28]

The other model is that there are receptors in plant cell membranes for fungal elicitors. This model is supported by many experimental results. For example, Yoshikawa et al.[29] demonstrated that membrane fractions prepared from soybean cotyledons contain a specific binding site for ^{14}C-labeled mycolaminaran, an intercellular β-1,3-glucan elicitor from *Phytophthora megasperma* f. sp. *glycinea*. Yoshikawa et al.[30] further demonstrated the presence of a receptor on soybean plasma membranes for elicitors released from the cell wall of *P. megasperma* f. sp. *glycinea* by β-1,3-glucanase activity of the host plant. That is, the total binding of a ^{14}C-labeled elicitor to isolated soybean membranes was inhibited by addition of an unlabeled elicitor in the dose-dependent manner. Several related carbohydrates without elicitor activity did not inhibit the binding. Thus, they concluded the existence of specific binding sites for released elicitors on soybean membranes. The same evidence was obtained by Schmidt and Ebel.[31] By using [^3H]1,3-glucan elicitors from *P. megasperma* f. sp. *glycinea*, they demonstrated the existence of a high-affinity elicitor binding site in membrane fractions of soybean roots. Competition studies with an [^3H]glucan elicitor and a number of polysaccharides indicated that only polysaccharides of the branched β-glucan type effectively displace the radiolabeled ligand from binding, and the displacing activity of glucans on elicitor binding corresponds closely to their respective ability to elicit phytoalexin production. Cosio et al.[32] proved that [^{125}I]glucan provides high sensitivity and lower detection limits for the binding assay, and found that the glucan binds to soybean protoplasts and microsomal fractions as well.

Because most of the fungal elicitors are high molecular weight compounds, it is hard to consider that these elicitors directly activate genes for resistance present in the nuclei through plant surface and nuclear membrane to chromosomes. From these standpoints, the latter model seems to be probable.

The detailed mechanisms on defense gene activation in the model are still obscure, but many available data suggest that some signal molecules are produced as the result of the binding of elicitor to receptor. Oguni et al.[33] and Kenn and Kennedy[34] reported that adenosine cyclic 3′,5′-phosphate (cAMP) induces phytoalexin and hypersensitive reaction in sweet potatoes and soybeans, respectively. Ethylene is reported to be produced in plants at an early stage of infection by pathogenic fungi or by treatment with phytoalexin elicitors.[35,36] The production of enzymes involved in phytoalexin biosynthesis, phenylalanine ammonia-lyase (PAL), and chalcone synthase, is also stimulated in plant tissues treated with ethylene.[37] An elicitor of ethylene production activates the biosynthesis of chitinase,[38] an antifungal enzyme, and hydroxyproline-rich glycoprotein[37] which is known to be a defense substance. These data support the hypothesis that ethylene may play some role as a mediator of events from elicitor binding sites to defense genes.

Contrary to positive evidence that ethylene may play a role as the signal molecule for defense gene activation, Gentile and Matta[39] reported that ethylene was induced in susceptible combinations of tomato and *Fusarium oxysporum* f. sp. *lycopersici*, corresponding to the occurrence of marked foliar wilting and basal leaf abscission; however, the differences in ethylene production were not observed in infected resistant plants or in plants inoculated with an avirulent form of *F. oxysporum*. Other evidence that the inhibitor or activator of ethylene biosynthesis does not affect the phytoalexin accumulation and disease resistance of plants also suggests that ethylene may not function as a signal messenger in a defense reaction.[35]

In studies using cultured plant cells,[40,41] elimination of Ca^{2+} from the culture media significantly reduces the response to the elicitor; and further, the direct measurement of ion concentrations in the culture media of elicitor-treated parsley cells revealed a decrease in Ca^{2+} and an increase in the K^{+} level within 2 min after the elicitor addition. The addition of Ca^{2+} ionophore, A 23187, stimulates phytoalexin accumulation in some cultured plant cells. These facts suggest the possibility of involvement of Ca^{2+} in elicitor-induced defense gene activation.

On the contrary, in experiments at the tissue level, Ca^{2+} or calmodulin may not be involved in the signal transduction in defense reaction[42,43] because chelators of Ca^{2+} and calmodulin inhibitors such as amytryptyline and chlorpromazine do not inhibit; however, the latter showed elicitor activity to the pea epicotyl.[43] This evidence shows that Ca^{2+} or calmodulin does not function in the signal transduction in defense reactions at the tissue or organ level.

Hargreaves et al.[44,45] found that heat stable elicitors are present in healthy bean tissue and named them endogenous or constitutive elicitors. They presented a scheme that these endogenous elicitors may be released by the stimuli of infection by pathogens, and function as signal molecules for phytoalexin synthesis. The same endogenous elicitors have been isolated from cell walls of tobacco and soybean plants, and it has been found that the elicitors were the hydrolysates of pectin having 10 to 15 galacturonic acid residues.[46,47] In this scheme, pectolytic enzymes secreted by pathogenic fungi or activated from plant tissue by infection hydrolyze the host cell wall pectin, and then the hydrolysate (endogenous elicitor) functions as the second messenger for defense gene activation.

Shiraishi et al.[43] have found that protein kinase seems to play an important role in the elicitation of defense reaction in the pea plant. That is, verapamil, a Ca^{2+} channel blocker, and K-252a, a strong inhibitor of protein kinase, inhibit pisatin accumulation in the pea epicotyl which has been treated with elicitor isolated from *Mycosphaerella pinodes*. Because $LaCl_3$ and 0,0′-bis(2-aminoethyl), ethyleneglycol-N,N,-N′N′ tetraacetic acid (EGTA) does not inhibit pisatin accumulation and also because verapamil inhibits pisatin accumulation even if it was applied 6 hr after the elicitor treatment, when the pisatin biosynthetic pathway has already been activated, it can hardly be considered that Ca^{2+} plays a role as the second messenger for signaling pisatin biosynthesis in pea tissue induced by an elicitor. On the other hand, K-252a inhibits pisatin accumulation when it is applied to a pea epicotyl only before the elicitor treatment. This fact suggests that signal transduction occurs very rapidly after the elicitor treatment, and further that protein kinase may play a key role in signal transduction for pisatin biosynthesis because K-252a inhibits markedly the in vitro phosphorylation of pea plasma membrane proteins.

As to the substrate of protein kinase in signal transduction in plants, almost nothing is known in contrast to what is known about animals in which phosphorylation of specific proteins is related to transduction of environmental, developmental, and metabolic signals.[48]

Kobayashi[49] using his system of barley and *Erysiphe pisi*, a nonhost resistance combination, recently found that rearrangement of the cytoplasmic strand, mainly composed of actin filaments, plays a very imporant role in the expression of the defense reaction. That is, the several cytoplasmic strands composed of actin filament were rearranged and appeared underneath the appressoria when inoculated. However, Oku et al.[50] found that K-252a completely inhibits the appearance of actin filaments. This is not direct evidence, but it is possible to assume that the phosphorylation of some proteins leading to the rearrangement of actin filament may be one of the processes for the expression of defense reactions not only in peas but also in barley.

Phosphorylation of lipid in plasma membranes also seems to be involved in the process of signal transduction between recognition of the elicitor by a putative receptor and defense gene activation including transmembrane signaling. That is, Chen and Boss[51] found that the rapid activation of inositol phospholipid kinases is induced in plasma membranes from cultured carrot cells pretreated with driselase and hemicellulase which are known to release endogenous elicitors from the cell wall. Toyoda et al.[52] demonstrated that the isolated pea plasma membrane has the activity to phosphorylate endogenous phosphatidylinositol (PI) and phosphatidylinositol-4-phosphate (PIP) even at 0°C within 5 sec, and the activity responds very rapidly and strongly to the elicitor prepared from spore germination fluid of *Mycosphaerella pinodes*.

Thus, the mechanism of elicitation of defense genes seems to be very complicated, but the clarification of the mechanism is an attractive subject for plant pathologists and biochemists. Because many defense genes are coordinately expressed after treatment with an elicitor, there might exist a plant nuclei a basic principle governing the activation of these defense-related genes. If one assumes the basic principle is a single dominant gene for resistance just as the vertical resistance gene in the race-cultivar relationship in cultivated plants,[53] it is possible to assume that the single dominant gene catches the final message which arrives via several steps of reactions after the binding of the elicitor with a receptor on the plasma membrane. As a result, another signal molecule may arise and activate many defense-related genes by interacting with *cis*- and *trans*-acting factors. The phosphorylation of some proteins and lipids may play a very important role in this process (Figure 1).

C. Gene Expression in Defense Response

There are many reports that the infection of plants by plant pathogens, especially incompatible pathogens, induces many kinds of isozymes which

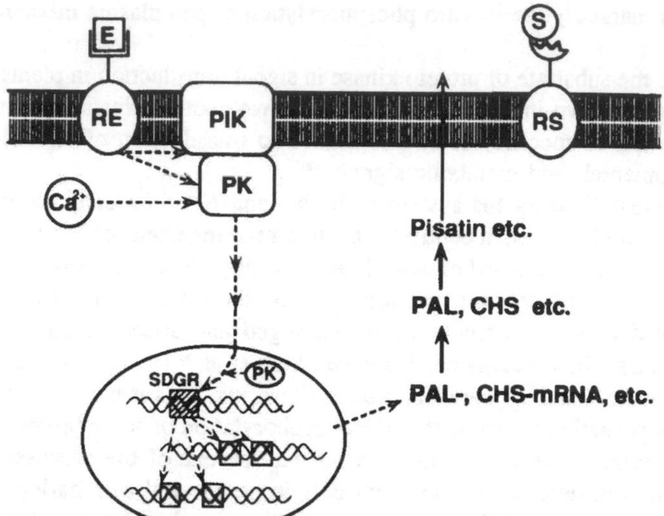

FIGURE 1. Schematic illustration of elicitation of defense reaction. E: Elicitor, RE: receptor for elicitor, S: suppressor, RS: receptor for suppressor, PK: protein kinase, PIK: phosphatidylinositol kinase, ▨: defense-related genes, SDRG ▩: single dominant gene for resistance.

are not present in uninfected plants; and these isozymes arise from net protein synthesis. Such experimental data show that genes encoding these isozymes are activated by infection. The same type of activation is also reported on genes encoding enzymes responsible for defense reactions.

For the convenience of the rapid and simultaneous distribution of an elicitor to plant cells, cultured plant cells have often been used for this type of experiment. Several examples are explained here.

As to the phytoalexin biosynthesis, activation of genes encoding enzymes involved in isoflavonoid phytoalexin has been detected by Northern blot hybridization with respective cDNAs as probes.

Isolated bean cells cultured in the dark were treated with an elicitor prepared from *Colletotrichum lindemuthianum*, and the mature mRNA accumulated in the treated cells were examined by Northern blot hybridization with cDNAs of phenylalanine ammonia-lyase (PAL), chalcone synthase (CHS), and chalcone isomerase (CHI) which are key enzymes for isoflavonoid phytoalexin biosynthesis. As the result, all mRNAs (PAL, CHS, and CHI) accumulated very rapidly, reached the maximum level 3 hr after treatment with an elicitor, and then decreased to the original level.[54–58]

Other than gene activation for phytoalexin biosynthesis, the mRNA encoding cinnamyl-alcohol dehydrogenase, an enzyme catalyzing the final step in a branch of phenyl propanoid synthesis specific to the synthesis of lignin monomers, accumulates in cultured bean cells very rapidly by the treatment with an elicitor from *C. lindemuthianum;* reaches the maximum after 1.5 hr; and then decreases to the original level 4 hr later. Southern blot analysis of bean genomic DNA indicates that the elicitor-induced cinnamyl-alcohol dehydrogenase is encoded by a single gene.[59] The transcription of genes encoding chitinase responsible for defense reaction by degrading the cell wall of pathogenic fungi is also activated very rapidly by the treatment with a fungal elicitor.[60]

The phenomena occurring in cultured plant cells do not necessarily explain the phenomena at the tissue and/or organ level. Habereder et al.[61] compared the induction of phenylalanine ammonia-lyase and chalcone synthase mRNA during fungal infection of soybean roots by *Phytophthora megasperma* f. sp. *glycinea* with that induced by soybean cell cultures by an elicitor. As a result, both mRNAs increased 1–2 hr after inoculation of the root with an incompatible race of the pathogenic fungus. In the compatible interactions, only a slight early enhancement of mRNA levels was found, and no further increase occurred until 9 hr after inoculation. The kinetics for the enhancement of soybean cell suspension cultures with a glucan elicitor prepared from the cell wall of *P. megasperma* were similar to that measured during the early stages of the resistant response of soybean roots.

Pea epicotyl tissues grown in the dark also accumulate mRNAs for phenylalanine ammonia lyase and chalcone synthase 1 hr after treatment with an elicitor prepared from the spore germination fluid of the pea pathogen, *Mycosphaerella pinodes.*[62]

Further, at the site of infection, expression patterns of several defense-related genes involved in phenylpropanoid biosynthesis in nonhost resistance were studied in situ RNA hybridization.[63] All mRNAs tested accumulate transiently and locally around the infection site of the primary leaf of parsley inoculated with *Phytophthora megasperma* f. sp. *glycinea,* a soybean pathogen, but is nonpathogenic to parsley. Phenylalanine ammonia-lyase mRNA was proved to accumulate around the site of penetration of young potato leaves infected by the late blight fungus, *Phytophthora infestans.*[64] This was confirmed by a combination of immunochemistry and in situ RNA-RNA hybridization. The accumulation was more rapid in incompatible interactions than in compatible ones of a selected potato cultivar, carrying resistance gene R_1 against appropriate races of the fungus.

The rapid activation of these gene expressions in plant cells after elicitor treatment or fungal attack indicates that in plant cells, very rapid steps may be involved in the signal transduction system from the recognition of microorganisms to the transcriptional activation of these genes.[58]

Enzymes involved in phenylpropanoid biosynthesis which are induced by a different kind of elicitors and those in unelicited bean cells were compared by in vitro translation using the mRNA-dependent rabbit reticulocyte lysate system and the polysomal-mRNA isolated from elicitor-treated and untreated cells.[65] According to chromatofocusing analysis, isoforms of PAL and CHS that were induced by culture filtrate and cell wall elicitors prepared from *C. lindemuthianum* showed similar patterns. However, these isoforms were different from those observed in unelicited cells.

As described in Chapter 2, hydroxyproline-rich glycoproteins (HRGPs) are believed to play roles in resistance of plants against pathogens as a structural barrier and as an agglutinin. A remarkable increase in HRGPs occurs not only during infection but also following wounding and elicitor treatment.[66,67] Several mRNAs that hybridize to genomic clones of HRGPs have been reported.[68,69] The response of a gene for an HRGP in suspension cultured cells of the tomato to an elicitor was less rapid and more prolonged than that observed in genes encoding enzymes of phytoalexin biosynthesis. These facts suggest the involvement of a secondary signal substance which originates indirectly from host cells.[58,69] HRGP mRNA also accumulates during race-cultivar specific interactions between bean hypocotyls and *Colletotrichum lindemuthianum*.[69]

An in vitro translation experiment using a wheat germ translation system with RNA from melon plants indicated several alterations of peptide patterns during infection by *Colletotrichum lindemuthianum*.[70] Further, cytosine-rich RNA which codes for proline-rich peptides was separated by oligo(dG)-cellulose chromatography of poly(A) RNA (mRNA) from melons after inoculation, and subjected to in vitro translation again. As a result, two peptides with molecular weights of 54,500 and 56,000 were obtained. These peptides were suggested to be precursors of the peptide moiety of HRGP (55,000). The time course of their production after inoculation in the in vitro translation system coincides with the in vivo accumulation of HRGP.

As to the gene for pathogenesis-related protein, the pattern of activation, structure, and genomic organization of *prp1* encoding PR1 elicited by infection by *Phytophthora infestans* or culture filtrate-elicitor was clarified in a potato cultivar carrying resistance gene *R1*.[71] That is, according to the runoff transcription assay with isolated nuclei from potato leaves, the gene *prp1* is activated very rapidly after the elicitor treatment; reached a maximum level of expression 1 hr later; and then decreased. The pattern is very similar to that of 4-coumarate-CoA lygase. The coding sequence of the *prp1* and the deduced amino acid sequence are strikingly similar to that of the 26-kDa heat shock protein from soybeans. In situ RNA hybridization assay revealed that the *prp1* transcript accumulated around the infection site.

As described above, genes encoding enzymes involved in phenylpropanoid biosynthesis are activated very rapidly after the elicitor treatment, but declined to the original level within several hours. The mechanism of this rapid decline of the accumulation of gene transcripts is not clear; however, Dixon[72] considered the possibility that the *trans*-cinnamic acid (the product of PAL) may play a role as a regulation signal in gene expression of the phenylpropanoid biosynthetic pathway, from the fact that *trans*-cinnamic acid inhibits the transcription of PAL and CHS genes in cultured bean cells.

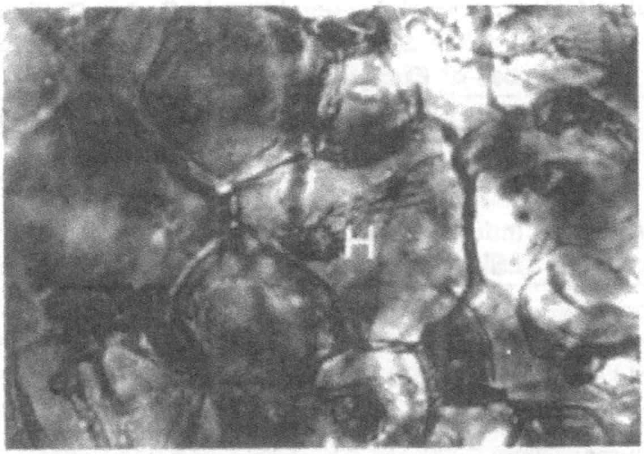

FIGURE 2. Characteristic haustorium (H) of *Erysiphe graminis* formed in the epidermal cell of melon leaf which was preliminarily inoculated with *Sphaerotheca fuliginea*.

II. SUPPRESSION OF DEFENSE REACTION OF HOST PLANTS BY COMPATIBLE PATHOGENS

A. Biological Evidence for Suppression of Defense Reaction

In spite of a variety of resistance mechanisms of host plants, pathogenic microorganisms of plants can establish infection on their own host plants. One possibility of doing this is the killing of host cells by deleterious toxins or enzymes, and then living on the dead cells in a saprophytic manner. However, this is not the case in obligate parasites because they cannot parasitize on dead cells.

Ouchi et al.[73,74] and Oku and Ouchi[75] found that the preliminary inoculation of the barley leaf with a compatible race of barley powdery mildew fungus (*Erysiphe graminis* f. sp. *hordei*) induced susceptibility not only to the originally incompatible race of the same fungus but also to the nonpathogenic powdery mildew fungi on barley; *Erysiphe graminis* f. sp. *tritici*, a wheat pathogen; or *Sphaerotheca fuliginea*, a melon pathogen. The same is true in melon leaves inoculated with *S. fuliginea*. That is, in melon leaf cells preliminarily inoculated with *S. fuliginea*, the barley powdery mildew fungus (*E. graminis*) forms characteristic globe-shaped haustorium (Figure 2), continues to grow, and finally forms abundant conidia on melon leaves. The conidia of the barley fungus thus produced on melon leaves were only pathogenic to barley but not to melon.

Inversely, the preliminary inoculation with incompatible races or *formae speciales* of *E. graminis* induces resistance to originally compatible races. For example, inoculation of barley leaves with a wheat pathogen, *E. graminis*

Table 1. Induction of Acceptability in Barley Cultivar H.E.S.4 to *Spaerotheca fuliginea* by Several Races of *Erysiphe graminis*

Inducer race	Inducer infectivity (ESH frequency[a] %)	Affinity indices of *S. fuliginea*: ESH frequency (%)	Affinity indices of *S. fuliginea*: SH length (μm)[b]
None		0	—
Hh4	33.6	48.6	315
Hr74	25.3	8.2	165
t2	3.6	1.7	140

Source: Guest.[75]

Note: S. fuliginea was inoculated 48 hr after the inoculation with an inducer race, and the affinity indices were determined after 48 hr of additional incubation.

[a] ESH frequency: frequency of spores elongating the secondary hyphae.
[b] SH length: length of the secondary hyphae.

f. sp. *tritici*, induces resistance to an originally compatible race of the barley pathogen.

Ouchi et al.[73] proposed the term "induced accessibility" to this type of susceptibility induction to distinguish the general concept of "induced susceptibility" which operates at the tissue or organ level, because the susceptibility induced by compatible pathogens occurs at the cellular level. The use of this terminology has been accepted by Kunoh et al.[76] who used the term "induced inaccessibility" instead of this type of induced resistance.

Table 1 shows the result of accessibility induction to *S. fuliginea* by several races and a forma specialis of *E. graminis*.[75] *S. fuliginea* is a nonpathogen of barley, but becomes established on barley leaves when the leaves are inoculated preliminarily with *E. graminis*. The infectivity and the hyphal growth of the challenger fungus, *S. fuliginea*, on barley leaves are positively correlated with the compatibility of the inducer fungi to barley. From these facts, it seems to be inappropriate to divide the compatibility of races or formae speciales as compatible and incompatible, because the compatibility of the fungus differs from fungus to fungus with a continuous manner. The term "highly compatible" and "less compatible" may be more appropriate, and we can estimate the degree of compatibility of races or formae speciales of a pathogen to their own host by measuring the degree of accessibility induction against the nonpathogenic challenger fungus.

These results suggest that the initial contact of an invading microorganism conditions the compatibility of host cells. Host cells express a series of defense

Table 2. Suppressors of Defense Reaction Found in Plant Pathogenic Fungi

Pathogenic fungus	Composition	Host plant	Ref.
Phytophthora infestans	Glucan, phosphoglucan	Potato	83
P. infestans	?	Tomato	84
Mycosphaerella pinodes	Glycopeptide	Pea	86
M. melonis	Glycopeptide?	Cucumber	86
M. ligulicola	Glycopeptide?	Chrysanthemum	86
Ascochyta rabiei	Glycoprotein	Chick pea	90
P. megasperma f. sp. *glycinea*	Mannan-glycoprotein	Soybean	91

reactions by contact with less compatible fungi; and as a result, the originally compatible pathogen cannot infect. The induction of accessibility in plant cells to the incompatible or nonpathogen by inoculation with a compatible pathogen indicates that the compatible pathogen has some mechanisms to suppress the defense response of its own host. The same kind of results were obtained by Tsuchiya and Hirata.[77] That is, the neighboring cells of the *E. graminis*-infected cells of barley became susceptible to many nonpathogenic powdery mildew fungi. Kunoh et al.[76,78] conducted more detailed experiments to determine the timing for induced susceptibility (accessibility) and resistance by inoculating *E. graminis* as a compatible pathogen and *E. pisi* as a nonpathogen on the same coleoptile cell of barley using a micromanipulator. Inoculation with *E. pisi* alone never infects the barley coleoptile cell, but penetration efficiency increased about 30% if the coleoptile cell was inoculated with *E. graminis* 60 min or more — earlier than the inoculation with *E. pisi*. Inversely, when *E. pisi* attempted penetration 60 min earlier than *E. graminis* on the same coleoptile cell, the penetration efficiency of *E. graminis* was reduced from 75 to 29%.

In the potato late blight disease, the inoculation with a compatible race of *Phytophthora infestans* suppressed the phytoalexin accumulation and the hypersensitive response.[79] The active mechanism to suppress these defense responses was considered to lead to susceptibility.[80]

B. Mechanism of Suppression of Defense Reaction

1. Suppressors of Defense Reaction Produced by Pathogenic Fungi

As described before, in several diseases the mechanism of host-parasite specificity at the race-cultivar level can be explained by specific elicitors. However, many pathogens are known to produce nonspecific elicitors, especially in diseases at the species-species level. In these cases, a mechanism other than the specific elicitor might determine the host-parasite specificity. Further, several plant pathogenic fungi have been known to produce substances which suppress the defense response of host cells (Table 2). These substances are called suppressors. A water-soluble fraction obtained from

Table 3. Infection and Colonization of *Mycosphaerella pinodes* on Various Leguminous Plants and Those of *Alternaria alternata* 15B as Affected by F5 Produced by *M. pinodes*

Higher plant	Degree of formation of intracellular hyphae[a]		
	M. pinodes	*Alternaria* 15B	15B[b] + F5[c]
Pisum sativum	4	0	4
Trifolium pratense	1	0	1
T. repens	0	0	0
Lotus corniculatus var. *Japonicus*	0	0	0
Millettia japonica	2	0	1
Lespedeza buergeri	2	0	2
L. bicolor	0	0	0
Vigna sinensis	0	0	0
Glycine max	1	0	1
Vicia faba	0–1	0	0
Arachis hypogaea	0	0	0

Note: Experiment was repeated three times but results were similar.

[a] Based on 0–4 rating where 0 = no formation and 4 = abundant formation.
[b] *Alternaria alternata* 15B.
[c] Partially purified suppressor.

mycelia of the compatible race of *Phytophthora infestans* suppressed hypersensitive reaction and phytoalexin production in potato disks, and the fraction isolated from the incompatible race was less active.[81] The component of the water-soluble fraction was composed of 17–24 glucose units, linked with β-(1–3) and β(1–6)-linkages.[82] Doke et al.[83] considered that the glucan plays a role as the determinant of specificity in the potato-*P. infestans* interaction. Storti et al.[84] also reported that substances released by germinating sporangia of *P. infestans*, but not by the mycelium, suppressed the hypersensitive reaction and phytoalexin production of the tomato.

A mannan-glycoprotein isolated from culture fluid of *Phytophthora megasperma* f. sp. *glycinea* suppresses glyceollin accumulation in soybeans induced by an elicitor from the same fungus.[85] This inhibition was reported to be race specific.

The pea pathogen, *Mycosphaerella pinodes*, secretes an elicitor and a suppressor for the biosynthesis of pisatin, a pea phytoalexin, into the spore germination fluid.[86,87] The elicitor was found in a high molecular weight fraction of the fluid and identified as a polysaccharide (ca. M_r 70,000). The suppressor was a mass of low molecular weight glycopeptides.

Alternaria alternata 15B, an avirulent isolate of the pear pathogen, and *Stemphylium sarcinaeforme*, a red clover pathogen, colonized on peas and formed conidia on the suppressor-treated leaves 2 weeks later. Further, as indicated in Table 3, *Alternaria* 15B could establish infection on five species

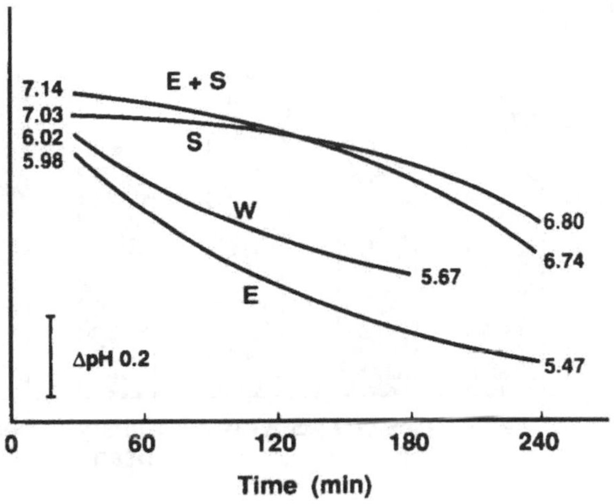

FIGURE 3. Effect of suppressor on the change of pH of droplets of elicitor solution or water placed on pea leaves. E: Elicitor, 100 μg/mL; S: suppressor, 50 μg/mL; W: water.

of leguminous plants to which *M. pinodes* was pathogenic in the presence of a suppressor, but not on other plant species. In other words, the host specificity of *M. pinodes* coincided completely with the specificity of biological activity of the suppressor. Thus, it was found that the suppressor might be a determinant for pathogenicity of *M. pinodes* by suppressing the general defense reaction.[86]

Suppressors do not cause any visible injury to the pea leaf, stem, and isolated protoplast prepared from the pea leaf; therefore they can hardly be "host-specific toxins." The kinetic analyses indicated that the mechanism of action of the suppressor is not competitive with the elicitor,[88] similar to the suppressor isolated from *Ascochyta rabiei*, a chick pea pathogen.[89,90]

Our recent results[88,91,92] indicate that the suppressor from *M. pinodes* inhibits the plasma membrane adenosinetriphosphatase (ATPase). Because the suppressor inhibits the decline of pH of the droplet of elicitor solution or water placed on the pea leaf (Figure 3), the suppressor may function by blocking the proton pump ATPase in the pea plasma membrane. This inhibitory activity of the suppressor is temporary on pea leaves because it disappeared within 2–3 hr after treatment.

Interesting is the fact that the specificity of suppressor is exhibited at the tissue level but not the in vitro level. As indicated in Table 3, activity of the suppressor prepared from *M. pinodes* showed specificity to plants to which this fungus is pathogenic, but not to nonhost plants. However, as shown in Figure 4, the inhibitory activity of the suppressor is nonspecific to ATPases of isolated membranes from the pea, bean, cowpea, soybean, and barley plants. In other words, the inhibitory activity of the suppressor to ATPase is

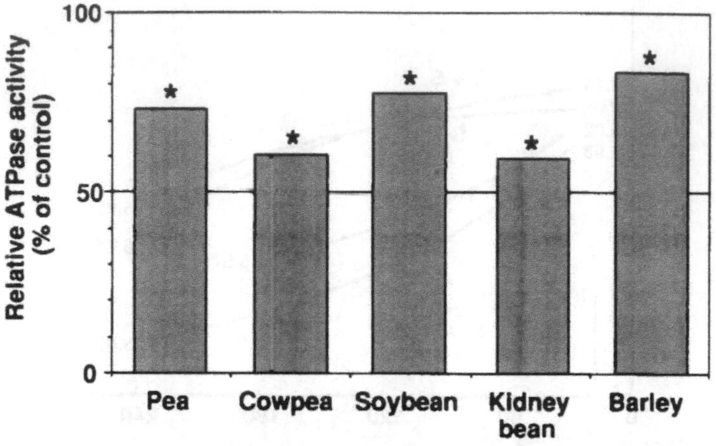

FIGURE 4. Effects of suppressor from *Mycosphaerella pinodes* on ATPase activities in plasma membrane fractions prepared from seedlings of peas and nonhost plants. The relative values of the ATPase activity in the presence of 50 μg/mL of suppressor are represented as percentages of the control value (addition of water). *Significantly different ($p \leq 0.05$) from the respective control value.

nonspecific in vitro. To elucidate this contradiction, the effect of the suppressor at the tissue level was examined using a lead precipitation procedure for ATPase under the electron microscope.[92] The principle of the method is reported by Hall et al.[93] and Moore et al.[94] That is, when ATP is added to living tissue, phosphate is released where ATPase is active and lead deposits are released when lead nitrate is added to the tissue. However, lead does not deposit in tissues where ATPase is inhibited because phosphate is not released from added ATP. As shown in Figure 5, vanadate inhibits ATPases nonspecifically of all plant species tested at the in vivo level. However, the suppressor from *Mycosphaerella pinodes* inhibits only ATPase of peas, but not of kidney beans, cowpeas, soybeans, and barley.[92] Namely, the suppressor is specific only in vivo, but not in vitro.

The specific inhibition of ATPase occurs at the host-parasite interface of the pea and *M. pinodes*. As shown in Figure 6, the inhibitory activity of ATPase is recognizable for 6 hr after inoculation, but not 9 hr later. As evident in these photos, no inhibition of ATPase occurred at the interface between peas and *Mycosphaerella ligulicola*, a nonpathogen of peas.

As described, the activity of lipid kinases (PI kinase and PIP kinase) in the isolated pea plasma membrane is activated by treatment with an elicitor. The suppressor inhibits activation of the lipid kinase activity induced by an elicitor.[52] It is likely that the mutual correlation may exist between the inhibition of ATPase and lipid kinase activity in the plasma membrane by a suppressor, though the detailed mechanism is unknown. However, it was reported that phospholipids might regulate ATPase activity in plasma membranes isolated from cultured carrot cells and sunflower hypocotyls.[95,96] Fur-

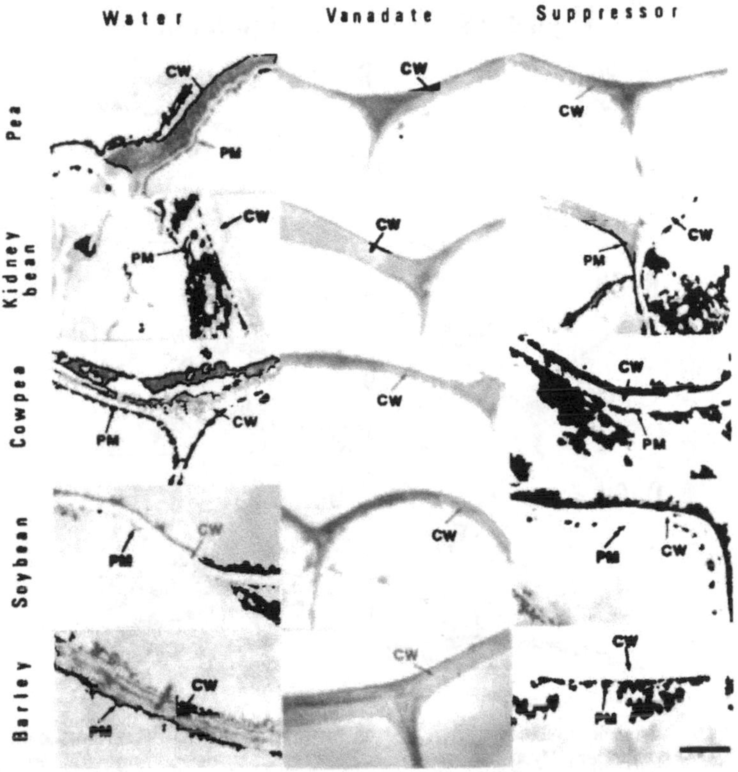

FIGURE 5. Effect of vanadate and suppressor from *Mycosphaerella pinodes* on plasma membrane ATPase activity of various plants at the tissue level examined by lead precipitation procedure. (From Shiraishi, T. et al., *Plant Cell Physiol.*, 32, 1070, 1991. With permission.)

ther, the elimination of lipids during the purification process of ATPase from a membrane fraction markedly reduced ATPase activity, and the addition of lipid recovered the activity.[97,98] These facts suggest that the ATPase activity might be regulated by circumferential lipid components. From these facts it is possible to consider that a suppressor inhibits the lipid kinase activity and as a result membrane ATPase activity is inhibited.

Inhibition of the plasma membrane ATPase might result in decline of the general cellular function and the signal transduction needed for elicitation of defense response delays. The pathogenic fungus may establish infection during this period.

Two kinds of suppressors were purified from spore germination fluid of *Mycosphaerella pinodes* by column chromatography and high-performance liquid chromatography (HPLC), and chemical structures were determined by amino acid analysis and ^{13}C- and ^{1}H-NMR[99,100] as shown in Figure 7. As seen in the figure, both suppressors were found to be mucin-type glycopeptides. The small suppressor of M_r 452 was named suppescin A and larger

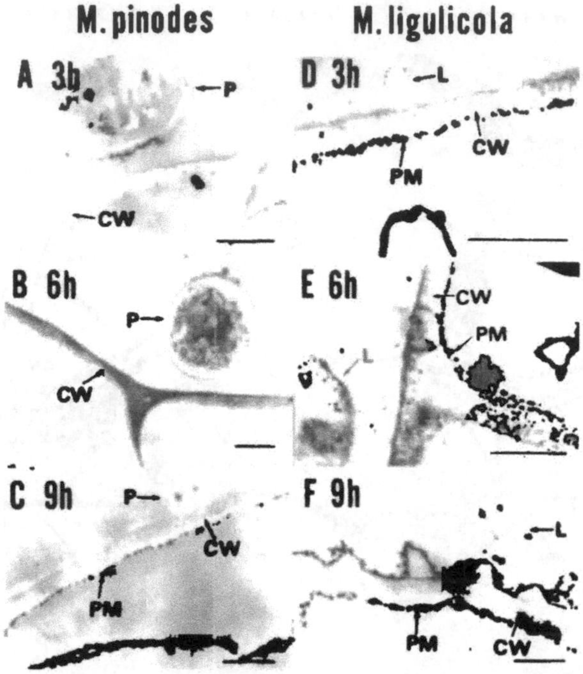

FIGURE 6. Specific inhibition of pea plasma membrane ATPase was caused by pea pathogen, *Mycosphaerella pinodes*, but not by nonpathogen, *M. ligulicola*. (From Shiraishi, T. et al., *Plant Cell Physiol.*, 32, 1073, 1991. With permission.)

α - GalNAc - O - Ser - Ser - Gly

Suppresсin A

β - Gal(1→4) - α - GalNAc - O - Ser - Ser - Gly - Asp - Glu - Thr

Suppresсin B

FIGURE 7. Structures of two suppressors, supprescin A and B, purified from spore germination fluid of the pea pathogen, *Mycosphaerella pinodes*.

one having M_r 959 was designated supprescin B. Since a full sequence of supprescin A is included in the molecule of supprescin B, both supprescins seem to be products of the same biosynthetic pathway; or supprescin A is produced by the splicing of supprescin B. Supprescin A and B suppressed pisatin biosynthesis in pea leaves that was induced by an elicitor at concentrations higher than 220 m*M* and 80 μ*M*, respectively. The synergistic or additive effect in the inhibition of pisatin biosynthesis was observed between them. Supprescin B inhibited 80% of the pea plasma membrane ATPase at a concentration of 320 μ*M*. Supprescin A did not inhibit ATPase activity even at a concentration of 2.2 m*M*. In the presence of 12.5 μ*M* of supprescin B, pea nonpathogen *Mycosphaerella ligulicola* could establish infection on pea leaves; but supprescin A did not do so at concentrations from 12.5 μ*M* to 5 m*M*.[100]

The essential part of supprescin B may be in the peptide moiety, because the treatment with proteolytic enzymes reduced the activity and chemically synthesized peptide moiety also inhibits ATPase activity. The activity of a chemically synthesized peptide to inhibit ATPase activity is low as compared to supprescin B.[99] Information from a spin coupling value of $_3J(NH-CH)$ and $_3J(C-O-CH)$, and of the nuclear Overhauser effect indicate that the peptide chains of these glycopeptides have an α-helix structure and that the sugar-peptide chains are in a "V" conformation. Further, the molecular dynamics simulation of supprescin B in water indicates that he base region of the V-shaped chain has an especially strong positive charge. This fact shows that the positively charged region may bind strongly with the negatively charged plasma membrane surface. This might be the reason why supprescin B is a more potent inhibitor of ATPase than of peptides.[99,101] Interesting is the fact that supprescin A does inhibit membrane ATPase in the presence of an elicitor. Thus, the actual role of these suppressors seems to be very complicated at the host-parasite interface.

2. Suppressors of Defense Reaction Produced by Host Plants

Some plants have been found to contain substances which suppress the defense reaction against pathogenic fungi. These substances are called endogenous suppressors.

Heath[102] reported that the preliminary inoculation with pathogenic rust fungi increased the number of haustorium formation of several nonpathogenic rust fungi around the haustorium of pathogenic fungi. Further, injection of an extract from susceptible rust-infected leaves of French beans increased haustorium formation by the bean rust fungus in cowpeas and the cowpea rust fungus in beans, namely, induced susceptibility to originally nonhost-nonpathogenic rust interactions. However, these extracts are not effective to the sunflower rust fungus in beans or cowpeas, or the copea rust fungus in sunflowers. In this experiment since the suppressing extracts for resistance are obtained from the intercellular fluid of rust-inoculated plants, therefore the active factor was treated as the endogenous suppressor; however, it seems to be possible that the suppressor for nonhost resistance may be secreted by rust fungi into the intercellular fluid of the host plants.

Intercellular fluid obtained from tomato leaves infected by *Cladosporium fulvum* suppresses the hypersensitive reaction and callose deposition in tomato leaves induced by nonspecific elicitors isolated from *C. fulvum*.[103] However, since the same suppressing activity has been detected in some, but not in all, intercellular fluids of healthy plants, the suppressor seems to be of host origin.

In this experiment, the authors suggest the possibility that suppression of the activity of a nonspecific elicitor is due to enzymatic degradation of the elicitor, because in vitro incubation of intercellular fluid with an elicitor decreases the elicitor activity and the suppressor is heat labile.

There are examples that a nonenzymatic suppressor for defense reaction is present in the host plant. Healthy pea leaves contain substances which suppress the pisatin biosynthesis induced by a fungal elicitor.[104,105] The molecular weights of the endogenous suppressors of peas are about 15,000 and 5000. The latter was identified as a glycopeptide. The treatment of pea leaves with these substances induced susceptibility to pea nonpathogens, *Mycosphaerella ligulicola* and *M. melonis*. These substances do not suppress the defense reaction in other leguminous plants such as kidney beans or cowpeas. One of these endogenous suppressors (M_r 5000) inhibits the plasma membrane ATPase similarly to the suppressor from *M. pinodes*.

Barley leaves contain a substance which enhances the infection frequency of incompatible powdery mildew fungi on barley.[106] The factor is a glycopeptide (M_r 3000–3500).[107] The factor also inhibits the plasma membrane ATPase of barley[108] in a similar manner to suppressors from *M. pinodes* and also as the endogenous suppressor from the pea plant. The glycopeptide from barley is composed of glycine, asparagine, glutamic acid, serine, and threonine in the peptide moiety; and acetylglucosamine, mannose, fucose, galactose, neuramic acid, and xylose in the sugar moiety. The infection-enhancing factor (endogenous suppressor) was also isolated from barley leaves infected with powdery mildew and was proved to be identical with that from healthy leaves. A larger amount of the glycopeptide diffused out from barley tissue when inoculated with a compatible race of powdery mildew fungus and immersed in water than with those inoculated with incompatible races. Therefore, the powdery mildew fungus, an obligate parasite, seems to have some mechanisms to make use of the infection-enhancing factor (endogenous suppressor) contained in host plants at the host-parasite interface. The infection of wheat by *Erysiphe graminis* f. sp. *tritici* or of pea leaves by *E. pisi* was not enhanced by the factor isolated from barley leaves. Thus, the factor seems to be responsible for the basic compatibility as suggested by Bushnell[109] and Heath[110,111] between barley and *E. graminis* at the species-species level.[109]

Elucidation of the mechanisms of action of these endogenous suppressors to compatible pathogens may be very important and useful in understanding the coevolutional processes between hosts and parasites.

3. Host-Specific Toxins as Suppressors of Defense Reaction at Early Stage of Infection

Alternaria alternata produces various kinds of host-specific toxins, and the specificity of the toxin to plant species or cultivars determines the host range of the pathogenic fungus.[112] The early important role of host-specific toxins in pathogenesis has been proved to be suppression of the defense reaction of the host plant.[113,114] For example, the spore germination fluid of *A. alternata* produces a nonspecific polysaccharide elicitor with a molecular weight of ca. 40,000 Da. This elicitor induces resistance in pear leaves by producing two kinds of infection-inhibiting factors (these factors have no antifungal activity and hence are called infection-inhibiting factors) to the invasion by a virulent *A. alternata*. However, the activity of the elicitor to produce infection inhibitors was completely suppressed when it was treated with AK toxin (host-specific toxin produced by *A. alternata* Japanese pear pathotype). Thus, an important role of AK toxin in the early stage of pathogenesis is to suppress the defense reaction in the host plant.[114] AK toxin does not inhibit the activity of membrane ATPase which was prepared from pear tissue.[115]

C. Suppression of Gene Expression Concerning Defense Reaction

In spite of many studies on gene expression for the host defense reaction as described above, few data are available on suppression of the gene activation by signal molecules produced by pathogenic fungi by which they can establish infection on their own hosts. Yamada et al.[62] elucidated that the suppressor prepared from spore germination fluid of *Mycosphaerella pinodes* (a pea pathogen) delays the expression of PAL and CHS genes encoding key enzymes for the pisatin biosynthetic pathway. Pea epicotyls grown in the dark have scarce amounts of PAL and CHS. However, treatment of the epicotyl with an elicitor prepared from *M. pinodes* induces the accumulation of PAL mRNA and CHS mRNA within 1 hr as described above. Under the concomitant presence of a suppressor with an elicitor, initiation of the transcriptional activation of these genes delays for 3 hr. A 3-hr delay of the activation of these genes results in a 6-hr delay of appearance of these enzymes and a 6–9-hr delay of pisatin accumulation in the pea epicotyl. Thus the suppressor of the pathogenic fungus suppresses the expression of genes for phytoalexin biosynthesis.

Cypers et al.[64] conducted in situ hybridization experiments of PAL mRNA with ^{32}P-labeled PAL antisense RNA as a probe at the infection site of potato leaves inoculated with compatible or incompatible races of the late blight fungus. As a result, they found that the remarkable accumulation of PAL mRNA in leaves inoculated with the compatible race delayed 3 hr compared with that inoculated with the incompatible race. The coincidence of a 3-hr delay in this result supports the concept that the suppressor produced by the pathogenic fungus plays a key role as the determinant of pathogenicity not only at the species-species level but also at the cultivar-race level.

III. DETOXIFICATION OF PHYTOALEXINS BY PATHOGENIC FUNGI

In some plant-parasite interactions, the detoxification of phytoalexins plays an important role in the pathogenicity of pathogenic fungi. According to

Cruickshank,[116] 5 fungal species or strains out of 50 tested were tolerant to pisatin, and all 5 were pea pathogens. Only 1 of 45 sensitive strains was a pea pathogen. Uehara[117] reported that a pea pathogen, *Ascochyta pisi*, metabolized pisatin to a less toxic product. According to Nonaka,[118] pea pathogens *Fusarium oxysporum* f. sp. *pisi* and *A. pisi* degrade pisatin but nonpathogens of peas do not; and he found positive correlations between pathogenicity to peas and the pisatin degrading ability of fungi.

The first step of pisatin degradation by many pea pathogens was shown to be the removal of the 3-*O*-methyl group to yield 6a-hydroxymaakiain. VanEtten[119,120] proved that the demethylated product was less toxic to many plant pathogenic fungi than pisatin. This was confirmed by Nonaka.[121]

Conclusive evidence that the phytoalexin degrading ability important in pathogenicity was obtained by transformation of genes for detoxification of pisatin into nonpathogenic fungi. That is, gene encoding pisatin demethylase was isolated from a virulent strain of *Nectria haematococca*[122] and was transformed into an avirulent strain of *N. haematococca* or the maize pathogen, *Cochliobolus heterostropus*. The transformed avirulent fungi were found to become virulent to peas.[123] These results indicate that, in addition to the importance of the phytoalexin degrading ability in pathogenicity of some pathogenic fungi, pisatin is the key compound responsible for resistance of the pea plant.

REFERENCES

1. Cruickshank, I. A. M. and Perrin, D. R., Studies on phytoalexins. IV. Pisatin: the effect of some factors on its formation in *Pisum sativum* L. and the significance of pisatin in disease resistance, *Aust. J. Biol. Sci.*, 16, 111, 1963.
2. Oku, H., Nakanishi, T., Shiraishi, T., and Ouchi, S., Phytoalexin induction by some agricultural fungicides and phytotoxic metabolites of pathogenic fungi, *Sci. Rep. Fac. Agric. Okayama Univ.*, 42, 17, 1973.
3. Reilly, J. J. and Klarman, W. L., The soybean phytoalexin, hydroxyphaseollin, induced by fungicides, *Phytopathology*, 62, 1113, 1972.
4. Schwoshau, M. E. and Hadwiger, L. E., Stimulation of pisatin production in *Pisum sativum* by actinomycin D and other compounds, *Arch. Biochem. Biophys.*, 26, 731, 1968.
5. Bridge, M. A. and Klarman, W. L., Soybean phytoalexin, hydroxyphaseollin, induced by ultraviolet irradiation, *Phytopathology*, 63, 606, 1973.
6. Cruickshank, I. A. M. and Perrin, D. R., The isolation and partial characterization of phaseollin-inducing activity from *Monilinia fructicola*, *Life Sci.*, 7, 449, 1968.
7. Keen, N. T., Specific elicitors of plant phytoalexin production: determinants of race specificity in pathogens?, *Science*, 187, 74, 1975.
8. Dervill, A. G. and Albersheim, P., Phytoalexins and their elicitors — defense against microbial infection in plants, *Annu. Rev. Plant Physiol.*, 35, 243, 1984.
9. Asada, Y., Induced lignification and elicitors, *Abstr. 5th ICPP*, Kyoto, 1988, 216.

10. Keen, N. T. and Yoshikawa, M., β-1,3-Glucanase from soybean releases elicitor-active carbohydrates from fungus cell wall, *Plant Physiol.*, 71, 460, 1983.
11. Yoshikawa, M., Matama, M., and Masago, H., Release of a soluble phytoalexin elicitor from mycelial wall of *Phytophthora megasperma* var. *sojae* by soybean tissues, *Plant Physiol.*, 67, 1032, 1981.
12. Yoshikawa, M. and Takeo, K., A structural model for the release of elicitors from cell wall of *Phytophthora megasperma* f. sp. *glycinea* by soybean β-1,3-endoglucanase, *Abstr. 5th ICPP*, Kyoto, 1988, 230.
13. Ayres, A. R., Valent, B., Ebel, J., and Albersheim, P., Host pathogen interactions. XI. Composition and structure of wall-released elicitor fractions, *Plant Physiol.*, 57, 766, 1976.
14. Cline, K., Wade, M., and Albersheim, P., Host-pathogen interactions. XV. Fungal glucans which elicit phytoalexin accumulation in soybeans also elicit the accumulation of phytoalexins in other plants, *Plant Physiol.*, 62, 918, 1978.
15. Yamamoto, Y., Oku, H., Shiraishi, T., Ouchi, S., and Koshizawa, K., Non-specific induction of pisatin and local resistance in pea leaves by elicitors from *Mycosphaerella pinodes, M. molonis,* and *M. ligulicola* and the effect of suppressor from *M. pinodes, J. Phytopathol.*, 117, 136, 1986.
16. Anderson, A. J., Studies on the structure and elicitor activity of fungal glucans, *Can. J. Bot.*, 58, 2343, 1980.
17. Anderson, A. J., Differences in the biochemical compositions and elicitor activity of extracellular components produced by three races of a fungal plant pathogen, *Colletotrichum lindemuthianum, Can. J. Microbiol.*, 26, 1473, 1980.
18. Tepper, C. S. and Anderson, A. J., Two cultivars of bean display a differential response to extracellular components from *Colletotrichum lindemuthianum, Physiol. Mol. Plant Pathol.*, 29, 411, 1986.
19. Keen, N. T. and Legrand, M., Surface glycoproteins: evidence that they may function as the race specific phytoalexin elicitors of *Phytophthora megasperma* f. sp. *glycinea, Physiol. Plant Pathol.*, 17, 175, 1980.
20. Daniel, S., Tiemann, K., Wittkampf, U., Bless, W., Hiderer, W., and Barz, W., Elicitor-induced changes in cell cultures of chickpea (*Cicer arientinum* L.) cultivars resistant and susceptible to *Ascochyta rabiei*. I. Investigation of enzyme activities involved in isoflavone and pterocarpan phytoalexin biosynthesis, *Planta*, 182, 270, 1990.
21. Gunia, W., Hinderer, W., Wittkampf, U., and Barz, W., Elicitor induction of cytochrome P-450 monoxygenases in cell suspension cultures of chickpea (*Cicer arientum* L.) and their involvement in pterocarpan phytoalexin biosynthesis, *Z. Naturforsch.*, 46c, 518, 1991.
22. Keen, N. T., Specific elicitors of plant phytoalexin production; determinants of race specificity in pathogens, *Science*, 187, 74, 1975.
23. Shiraishi, T., Oku, H., Yamashita, M., and Ouchi, S., Elicitor and suppressor of pisatin induction in spore germination fluid of pea pathogen, *Mycosphaerella pinodes, Ann. Phytopathol. Soc. Jpn.*, 44, 659, 1978.
24. Hadwiger, L. A. and Schwochau, M. E., Host resistance: an induction hypothesis, *Phytopathology*, 59, 223, 1969.

25. Hadwiger, L. A. and Schwochau, M. E., Specificity of deoxyribonucleic acid interacting compounds in the control of phenylalanine ammonia-lyase and pisatin levels, *Plant Physiol.*, 47, 346, 1971.
26. Hadwiger, L. A., Jafri, A., vonBroembsen, S., and Eddy, R., Jr., Mode of pisatin induction, increased template activity and dye-binding capacity of chromatin isolated from polypeptide-treated pea pods, *Plant Physiol.*, 53, 52, 1971.
27. Bridge, M. A. and Klarman, W. L., Soybean phytoalexin, hydroxyphaseollin induced by ultra violet irradiation, *Phytopathology*, 63, 606, 1973.
28. Hadwiger, L. A., Beckman, J. M., and Adams, M. J., Localization of fungal components in the pea-*Fusarium* interaction detected immunochemically with anti-chitosan and antifungal cell wall antisera, *Plant Physiol.*, 67, 170, 1981.
29. Yoshikawa, M., Keen, N. T., and Wang, M. C., A receptor on soybean membranes for a fungal elicitor of phytoalexin accumulation, *Plant Physiol.*, 73, 49, 1983.
30. Yoshikawa, M., Sugimoto, K., and Masago, H., Receptor on soybean membranes for the elicitors released from cell walls of *Phytophthora megasperma* f. sp. *glycinea* by soybean β-1,3-glucanase, *Abstr. 5th ICPP*, Kyoto, 1988, 216.
31. Schmidt, W. E. and Ebel, J., Specific binding of a fungal glucan phytoalexin elicitor to membrane fractions from soybean *Glycine* max, *Proc. Natl. Acad. Sci. U.S.A.*, 84, 4117, 1987.
32. Cosio, E. G., Popperl, H., Schmidt, E., and Ebel, J., High-affinity binding of fungal glucan fragments to soybean (*Glycine* max L.) microsomal fractions and protoplasts, *Eur. J. Biochem.*, 175, 309, 1988.
33. Oguni, I., Suzuki, K., and Uritani, I., Terpenoid induction in sweet potato roots by cyclic-3⁻,5⁻-adenosine monophosphate, *Agric. Biol. Chem.*, 40, 1251, 1976.
34. Keen, N. T. and Kennedy, B. W., Hydroxyphaseollin and related isoflavonoids in the hypersensitive resistance reaction of soybeans to *Pseudomonas glycinea*, *Physiol. Plant Pathol.*, 4, 173, 1974.
35. Paradies, I., Konze, J. R., Elstner, E. F., and Paxton, J., Ethylene: indicator but not inducer of phytoalexin synthesis in soybean, *Plant Physiol.*, 66, 1106, 1980.
36. Toppan, A. and Esquerré-Tugayé, M. T., Cell surface in plant-microorganism interactions. IV. Fungal glycopeptides which elicit the synthesis of ethylene in plant, *Plant Physiol.*, 75, 1133, 1984.
37. Ecker, J. R. and Davis, R. W., Plant defense genes are regulated by ethylene, *Proc. Natl. Acad. Sci. U.S.A.*, 84, 5202, 1987.
38. Roby, D., Toppan, A., and Esquerré-Tugayé, M. T., Cell surface in plant-microorganism interactions. VI. Elicitors of ethylene from *Colletotrichum lindemuthianum* trigger chitinase activity in melon plants, *Plant Physiol.*, 81, 228, 1986.
39. Gentile, A. and Matta, A., Production of and some effects of ethylene in relation to *Fusarium* wilt of tomato, *Physiol. Plant Pathol.*, 5, 27, 1975.
40. Staeb, A. R. and Ebel, J., Effects of Ca^{2+} on phytoalexin induction by fungal elicitor in soybean cells, *Arch. Biochem. Biophys.*, 257, 416, 1987.
41. Scheel, D. and Parker, J. E., Elicitor recognition and signal transduction in plant defense gene activation, *Z. Naturforsch.*, 45c, 569, 1990.

42. Kendra, D. F. and Hadwiger, L. A., Calcium and calmodulin may not regulate the disease resistance and pisatin formation responses of *Pisum sativum* to chitosan or *Fusarium solani*, *Physiol. Mol. Plant Pathol.*, 31, 337, 1987.
43. Shiraishi, T., Hori, N., Yamada, T., and Oku, H., Suppression of pisatin accumulation by an inhibitor of protein kinase, *Ann. Phytopathol. Soc. Jpn.*, 56, 261, 1990.
44. Hargreaves, J. A. and Baily, J. A., Phytoalexin production by hypocotyls of *Phaseolus vulgaris* in response to constitutive metabolites by damaged bean cells, *Physiol. Plant Pathol.*, 13, 89, 1987.
45. Hargreaves, J. A. and Selby, C., Phytoalexin formation in cell suspension of *Phaseolus vulgaris* in response to an extract of bean hypocotyls, *Phytochemistry*, 17, 1099, 1978.
46. Jin, D. F. and West, C. A., Characteristics of galacturonic acid oligomers as elicitors of casbene synthetase activity in castor been seedlings, *Plant Physiol.*, 74, 989, 1984.
47. Nathnagel, E. A., McNeil, A. M., Albersheim, P., and Dell, A., Host-pathogen interactions. XXII. A galacturonic acid oligosaccharide from plant cell wall elicits phytoalexins, *Plant Physiol.*, 72, 916, 1983.
48. Edelman, A. M., Blumenthal, D. K., and Krebs, E. G., Protein serine/threonine kinases, *Annu. Rev. Biochem.*, 56, 569, 1987.
49. Kobayashi, I., Induction of inaccessibility and the recognition of non-pathogen, *Erysiphe pisi* by barley leaf-sheath cell, Ph.D. thesis, submitted to Okayama University, 1991, 35.
50. Oku, H., Kobayashi, I., and Kunoh, H., unpublished data.
51. Chen, Q. and Boss, W. F., Short-term treatment with cell wall degrading enzymes increases the activity of the inositol phospholipid kinases and the vanadate-sensitive ATPase activity, *Plant Physiol.*, 94, 1820, 1990.
52. Toyoda, K., Shiraishi, T., Yoshioka, H., Yamada, T., Ichinose, Y., and Oku, H., Regulation of polyphosphoinositide metabolism in pea plasma membranes by elicitor and suppressor from a pea pathogen, *Mycosphaerella pinodes*, *Plant Cell Physiol.*, 33, 445, 1992.
53. Flor, H. H., The complementary genetic systems in flax and flax rust, *Adv. Genet.*, 8, 29, 1956.
54. Hahlbrock, K. and Scheel, D., Physiology and molecular biology of phenylpropanoid metabolism, *Annu. Rev. Plant Physiol.*, 40, 347, 1989.
55. Edwards, K., Cramer, C. L., Bolwell, G. P., Dixon, R. A., Schuch, W., and Lamb, C. J., Rapid transient induction of phenylalanine ammonia-lyase mRNA in elicitor-treated bean cells, *Proc. Natl. Acad. Sci. U.S.A.*, 82, 6731, 1985.
56. Mehdy, M. C. and Lamb, C. J., Chalcone isomerase cDNA cloning and mRNA induction by fungal elicitor, wounding and infection, *EMBO J.*, 6, 1527, 1987.
57. Ryder, T. B., Gramer, C. J., Bell, J. N., Robins, M. P., Dixon, R. A., and Lamb, C. J., Elicitor rapidly induce chalcone synthase mRNA in *Phaseollus vulgaris* cells at the onset of the phytoalexin defense response, *Proc. Natl. Acad. Sci. U.S.A.*, 81, 5724, 1984.
58. Lamb, C. J., Lawson, M. A., Dron, M., and Dixon, R. A., Signal and transduction mechanism for activation of plant defenses against microbial attack, *Cell*, 56, 215, 1989.

59. Walter, M. H., Grima-Pettenati, J., Grand, C., Boudet, A. M., and Lamb, C. J., Cinnamyl-alcohol dehydrogenase, a molecular marker specific for lignin synthesis: cDNA cloning and mRNA induction by fungal elicitor, *Proc. Natl. Acad. Sci. U.S.A.*, 85, 5546, 1988.
60. Hedrick, S. A., Bell, J. N., Bollar, T., and Lamb, C. J., Chitinase cDNA cloning and mRNA induction by fungal elicitor, wounding and infection, *Plant Physiol.*, 86, 182, 1988.
61. Habereder, H., Schroeder, G., and Ebel, J., Rapid induction of phenylalanine ammonia-lyase and chalcone synthase mRNAs during fungus infection of soybean; (*Clycine max* L.) roots or elicitor treatment of soybean cell cultures at the onset of phytoalexin synthesis, *Planta*, 177, 58, 1989.
62. Yamada, T., Hashimoto, H., Shiraishi, T., and Oku, H., Suppression of pisatin, phenylalanine ammonia-lyase mRNA, and chalcone synthase mRNA accumulation by a putative pathogenicity factor from the fungus *Mycosphaerella pinodes*, *Mol. Plant-Microbe Interact.*, 2, 256, 1989.
63. Schmelzer, E., Krugerl-Lebus, S., and Hahlbrock, K., Temporal and spatial patterns of gene expression around sites of attempted fungal infection in parsley leaves, *Plant Cell*, 1, 993, 1989.
64. Cypers, B., Schmelzer, E., and Hahlbrock, K., In situ localization of rapidly accumulated phenylalanine ammonia-lyase mRNA around penetration sites of *Phytophthora infestans* in potato leaves, *Mol. Plant-Microbe Interact.*, 1, 157, 1988.
65. Hamadan, A. M. S. and Dixon, R. A., Differential patterns of protein synthesis in bean cells exposed to elicitors from *Colletotrichum lindemuthianum*, *Physiol. Mol. Plant Pathol.*, 31, 105, 1987.
66. Esquerré-Tugayé, M. T., Mazeau, D., Toppan, A., and Roby, D., Elicitation via l'ethylene, de la synthese de glycoproteins paraietales associees a la defense des plantes, *Ann. Phytopathol.*, 12, 403, 1980.
67. Roby, D., Toppan, A., and Esquerré-Tugayé, M. T., Cell surfaces in plant microorganism interactions. V. Elicitor of fungal and plant origin trigger the synthesis of ethylene and of cell wall hydroxyproline-rich glycoproteins in plants, *Plant Physiol.*, 77, 700, 1985.
68. Chen, J. and Varner, J. E., An extracellular matrix protein in plants: characterization of a genomic clone for carrot extensin, *EMBO J.*, 4, 2145, 1985.
69. Showalter, A. M., Bell, J. N., Carmer, C. L., Baily, J. A., Varner, J. E., and Lamb, C. J., Accumulation of hydroxyproline-rich glycoprotein mRNA in response to fungal elicitor and infection, *Proc. Natl. Acad. Sci. U.S.A.*, 82, 6551, 1985.
70. Rumeau, D., Mazau, D., and Esquerré-Tugayé, M. T., Cytosine-rich RNAs from infected melon plants and their in vitro translation products, *Physiol. Mol. Plant Pathol.*, 31, 305, 1987.
71. Taylor, J. L., Fritzmeier, K., Hauser, I., Kombrink, E., Rohwer, F., Schroeder, M., Strittmatter, G., and Hahlbrock, K., Structural analysis and activation of fungal infection of a gene encoding a pathogenesis-related protein in potato, *Mol. Plant-Microbe Interact.*, 3, 72, 1990.
72. Dixon, R. A., The phytoalexin response: elicitation, signalling and the control of host gene expression, *Bot. Rev.*, 61, 239, 1986.

73. Ouchi, S., Oku, H., Hibino, C., and Akiyama, I., Induction of accessibility and resistance in leaves of barley by some races of *Erysiphe graminis*, *Phytopathol. Z.*, 79, 24, 1974.
74. Ouchi, S., Oku, H., Hibino, C., and Akiyama, I., Induction of accessibility to a non-pathogen by preliminary inoculation with a pathogen, *Phytopathol. Z.*, 79, 142, 1974.
75. Oku, H. and Ouchi, S., Host plant accessibility to pathogens, *Rev. Plant Prot. Res.*, 9, 58, 1976.
76. Kunoh, H., Katsuragawa, N., Yamaoka, N., and Hayashimoto, A., Induced accessibility and enhanced inaccessibility at the cellular level in barley coleoptile. III. Timing and localization of enhanced inaccessibility in a single coleoptile cell and its transfer to an adjacent cell, *Physiol. Mol. Plant Pathol.*, 33, 181, 1988.
77. Tsuchiya, K. and Hirata, K., Growth of various powdery mildew fungi on the barley leaves infected preliminarily with the barley powdery mildew fungus, *Ann. Phytopathol. Soc. Jpn.*, 39, 396, 1973.
78. Kunoh, H., Hashimoto, A., Harumi, M., and Ishizaki, H., Induced susceptibility and enhanced resistance at the cellular level in barley coleoptile. I. The significance of timing of fungal invasion, *Physiol. Plant Pathol.*, 27, 43, 1985.
79. Verns, J. L. and Kuć, J., Suppression of rishitin and phytuberin accumulation and hypersensitive response in potato by compatible races of *Phytophthora infestans*, *Phytopathology*, 61, 178, 1971.
80. Kuć, J., Phytoalexins, *Annu. Rev. Phytopathol.*, 10, 207, 1972.
81. Garas, N. A., Doke, N., and Kuć, J., Suppression of hypersensitive reaction in potato tubers by mycelial components from *Phytophthora infestans*, *Physiol. Plant Pathol.*, 15, 117, 1979.
82. Kuć, J., Henfling, J., Garas, N., and Doke, N., Control of terpenoid metabolism in potato-*Phytophthora infestans* interaction, *J. Food Prot.*, 42, 508, 1979.
83. Doke, N., Garas, N. A., and Kuć, J., Effect of host hypersensitivity suppressors released during the germination of *Phytophthora infestans* cystospore, *Phytopathology*, 70, 35, 1980.
84. Storti, E., Pelucchini, D., Tegli, S., and Scala, A., A potential defense mechanism of tomato against the late blight disease is suppressed by germinating sporangia-derived substances from *Phytophthora infestans*, *J. Phytopathol.*, 121, 275, 1988.
85. Ziegler, E. and Pontzen, R., Specific inhibition of glucan-elicited glyceollin accumulation in soybeans by an extracellular mannan-glycoprotein of *Phytophthora megasperma* f. sp. *glycinea*, *Physiol. Plant Pathol.*, 20, 321, 1982.
86. Oku, H., Shiraishi, T., Ouchi, S., Ishiura, M., and Matsueda, R., A new determinant of pathogenicity in plant disease, *Naturwissenschaften*, 67, 310, 1980.
87. Oku, H., Shiraishi, T., and Ouchi, S., Role of specific suppressors in pathogenesis of *Mycosphaerella* species, in *Molecular Determinants of Plant Diseases*, Nishimura, S., Vance, C. P., and Doke, N., Eds., Japan Scientific Press, Tokyo/Springer-Verlag, Berlin, 1987, 145.

88. Shiraishi, T., Yamada, T., Oku, H., and Yoshioka, H., Suppressor production as a key factor for fungal pathogenesis, in *Molecular Strategies of Pathogen and Host Plants*, Patil, S., Ouchi, S., Mills, D., and Vance, C., Eds., Springer-Verlag, New York, 1991, 151.
89. Kessmann, H. and Barz, W., Elicitation and suppression of phytoalexin and isoflavone accumulation in cotyledons of *Cicer arientinum* L. as caused by wounding and by polymeric components from the fungus *Ascochyta rabiei*, *J. Phytopathol.*, 117, 321, 1986.
90. Barz, W., personal communication, 1988.
91. Yoshioka, H., Shiraishi, T., Yamada, T., Ichinose, Y., and Oku, H., Suppression of pisatin production and ATPase activity in pea plasma membranes by orthovanadate, verapamil and a suppressor from *Mycosphaerella pinodes*, *Plant Cell Physiol.*, 31, 1139, 1990.
92. Shiraishi, T., Araki, M., Yoshioka, H., Kobayashi, I., Yamada, T., Ichinose, Y., Kunoh, H., and Oku, H., Inhibition of ATPase activity in pea plasma membranes in situ by a suppressor from a pea pathogen, *Mycosphaerella pinodes*, *Plant Cell Physiol.*, 32, 1067, 1991.
93. Hall, J., Browning, A. J., and Harvey, D. M. R., The validity of the lead precipitation technique for the localization of ATPase activity in plant cells, *Protoplasma*, 104, 193, 1980.
94. Moore, R., McClelen, C. E., and Smith, H. S., Phosphatases, in *Handbook of Plant Cytochemistry*, Vol. 1, Vaughn, K. C., Ed., CRC Press, Boca Raton, FL, 1987, 37.
95. Chen, Q. and Boss, W. F., Neomycin inhibits the phosphatidylinositol monophosphate and phosphatidylinositol biphosphate stimulation of plasma membrane ATPase activity, *Plant Physiol.*, 96, 340, 1991.
96. Memon, A. R., Chen, Q., and Boss, W. F., Inositol phospholipids activate plasma membrane ATPase in plants, *Biochem. Biophys. Res. Commun.*, 162, 1295, 1989.
97. Kasamo, K. and Nouchi, I., The role of phospholipids in plasma membrane ATPase activity in *Vigna radiata* L. (mung bean) roots and hypocotyls, *Plant Physiol.*, 83, 323, 1987.
98. Serrano, R., Montesinos, C., and Sanchez, J., Lipid requirements of the plasma membrane ATPase from oat roots and yeast, *Plant Sci.*, 56, 117, 1988.
99. Saitoh, K., Tahara, M., Kato, T., Tanaka, M., Kim, H., Shiraishi, T., Yamada, T., and Oku, H., Structural analysis of suppressors secreted from a pea pathogen, *Mycosphaerella pinodes*, by NMR spectroscopy and molecular dynamics, *Program & Abstr. IS-MPMI*, Seattle, 1992, Poster No. 310.
100. Shiraishi, T., Saitoh, K., Kim, H. M., Kato, T., Tahara, M., Oku, H., Yamada, T., and Ichinose, Y., Two suppressors, suppresscin A and B, secreted by a pea pathogen, *Mycosphaerella pinodes*, *Plant Cell Physiol.*, 33, 663, 1992.
101. Oku, H., Shiraishi, T., Saitoh, K., and Tahara, M., Structure and mode of action of suppressors, pathogenicity factors of pea pathogen, *Mycosphaerella pinodes*, *Abstr. 2nd EFPP Conf.*, Strasbourg, 1992.
102. Heath, M. C., Effect of injection by compatible species or injection of tissue extracts on susceptibility of nonhost plants to rust fungi, *Phytopathology*, 70, 356, 1980.

103. Peever, T. L. and Higgins, V. J., Suppression of the activity of non-specific elicitor from *Cladosporium fulvum* by intercellular fluids from tomato leaves, *Physiol. Mol. Plant Pathol.*, 34, 471, 1989.
104. Hori, N., Shiraishi, T., Yamada, T., and Oku, H., The role of endogenous suppressors in phytoalexin production by pea plant, *Abstr. 5th ICPP*, Kyoto, 1988, 231.
105. Shiraishi, T., Nasu, K., Yamada, T., Ichinose, Y., and Oku, H., Suppression of defense reaction and accessibility induction in pea by substances from healthy pea leaves, in *Molecular Strategies of Pathogen and Host Plants*, Patil, S., Ouchi, S., Mills, K. D., and Vance, C., Eds., Springer-Verlag, New York, 1991, 252.
106. Shiraishi, T., Miyazaki, T., Yamada, T., and Oku, H., Infection enhancing factor for *Erysiphe graminis* prepared from healthy barley seedlings, *Ann. Phytopathol. Soc. Jpn.*, 55, 357, 1989.
107. Oku, H., Shiraishi, T., Miyazaki, T., Yamada, T., and Ichinose, Y., Infection enhancing factor in barley, a substance possibly responsible for basic compatibility with *Erysiphe graminis*, in *Molecular Strategies of Pathogen and Host Plants*, Patil, S., Ouchi, S., Mills, D., and Vance, C., Eds., Springer-Verlag, New York, 1991, 253.
108. Nasu, K., Shiraishi, T., Yoshioka, H., Hori, N., Ichinose, Y., Yamada, T., and Oku, H., An endogenous suppressor of the defense response in *Pisum sativum*, *Plant Cell Physiol.*, 33, 617, 1992.
109. Bushnell, W. R., The nature of basic compatibility between pistil-pollen and host-parasite interaction, in *Recognition and Specificity in Host-Parasite Interaction*, Daily, J. M. and Uritani, I., Eds., Japan Scientific Society Press, Tokyo/University Park Press, Baltimore, MD, 1979, 211.
110. Heath, M. C., A general concept of host-parasite specificity, *Phytopathology*, 71, 1121, 1981.
111. Heath, M. C., Evolution of plant resistance and susceptibility to fungal invaders, *Can. J. Plant Pathol.*, 9, 389, 1987.
112. Nishimura, S., Kohmoto, K., Otani, H., Ramachandran, P., and Tamura, F., Pathological and epidemiological aspects of *Alternaria alternata* infection depending on a host-specific toxin, in *Plant Infection*, Asada, Y., Bushnell, W. R., Ouchi, S., and Vance, C. P., Eds., Japan Scientific Society Press, Tokyo/Springer-Verlag, Berlin, 1982, 199.
113. Hayami, N., Otani, H., Nishimura, S., and Kohmoto, K., Induced resistance in pea leaves by spore germination fluid of nonpathogens to *Alternaria alternata* Japanese pear pathotype and suppression of the induction by AK-toxin, *J. Fac. Agric. Tottori Univ.*, 17, 9, 1982.
114. Otani, H., Kohmoto, K., Kodama, M., and Nishimura, S., Suppression of resistance by toxins, *Abstr. 5th ICPP*, Kyoto, 1988, 220.
115. Otani, H., Kohmoto, K., and Nishimura, S., Action sites for AK-toxin produced by the Japanese pear pathotype of *Alternaria alternata*, in *Host-Specific Toxins: Recognition and Specificity Factors in Plant Disease*, Kohmoto, K. and Durbin, R. D., Eds., Tottori University Press, Tottori, 1989, 107.
116. Cruickshank, I. A. M., Studies on phytoalexins. IV. The antimicrobial spectrum of pisatin, *Aust. J. Biol. Sci.*, 15, 147, 1962.
117. Uehara, K., Relationship between host specificity of pathogen and phytoalexin, *Ann. Phytopathol. Soc. Jpn.*, 29, 103, 1964.

118. Nonaka, F., Inactivation of pisatin by pathogenic fungi, *Agric. Bull. Saga Univ.*, 24, 109, 1967.
119. VanEtten, H. D., Pueppke, S. G., and Kelsey, T. C., 3,6a-Dihydroxy-8,9-methylenedioxypterocarpan as a metabolite of pisatin produced by *Fusarium solani* f. sp. *pisi*, *Phytochemistry*, 14, 1103, 1975.
120. VanEtten, H. D. and Pueppke, S. G., Isoflavonoid phytoalexins, in *Biochemical Aspects of Plant Parasitic Relationship*, Friend, J. and Threeifall, D., Eds., *Annu. Proc. Phytochem. Soc.*, 239, 13, 1976.
121. Nonaka, F., Studies on the mechanism of plant disease resistance, especially on the phytoalexins, *Ann. Phytopathol. Soc. Jpn.*, 47, 284, 1981.
122. VanEtten, H. D., Matthews, D. E., and Mathews, P. S., Phytoalexin detoxification: importance for pathogenicity and practical implications, *Annu. Rev. Phytopathol.*, 27, 143, 1989.
123. Yorder, O. C., Altered virulence in recombinant fungal pathogens, *Abstr. 5th ICPP*, Kyoto, 1988, 226.

CHAPTER 4

Disease Control Agents Based on Knowledge of Pathogenicity and Disease Resistance

I. DISEASE OF CULTIVATED PLANTS AND FUNGICIDES

A. Conception of Pathogenicity

In the previous chapters, the mechanisms of pathogenesis in plant disease have been described; and this knowledge will be the basis for developing control measures for diseases, but it seems to be impossible to know all of the complexities of natural phenomena in detail. That is, plant diseases are not only caused by the ability of pathogens, but also are caused as the result of host-parasite interaction; and environmental conditions largely affect the outbreak of diseases (Figure 1). Therefore, the analyses of these conditions are also important in setting up disease control programs.

However, it is impossible to control plant disease by ecological, cultural, and epidemiological procedures, because of the abilities of pathogens such as vigorous multiplication, adaptability to environmental conditions, excellent strategies for preservation of species, and so on; therefore, a major part of the control of crop diseases has to be dependent on agrochemicals. In addition, cultivated plants are in one sense artificial plants, which have been improved by breeding to obtain higher yield, higher nutrition for mankind, and easier consumption. These crops are exclusively cultivated in mass in the field. In other words, agriculture is an industry of mass production of artificial plants. Therefore, it is natural to consider that these artificial plants are easily attacked

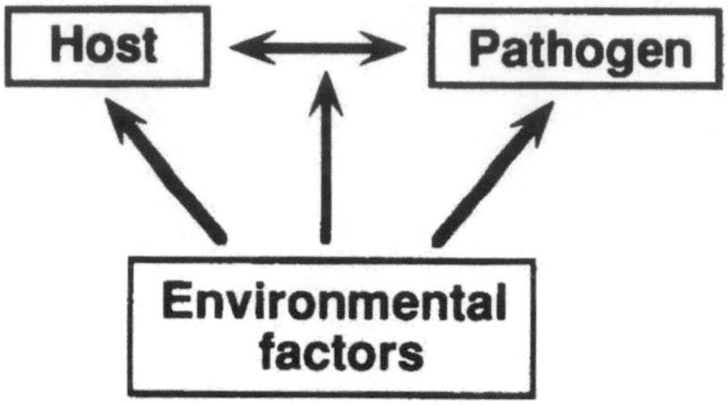

FIGURE 1. Diagram of outbreak of plant disease.

by many pathogens and insects. That is why we need agrochemicals to protect crop plants from diseases.

1. Fungicides and Environmental Pollution

The development of culture and economic status of human society is supported by an adequate, stable supply of good quality food materials. The main cause of the instability of food supply in ancient times was the destruction of crop plants by harmful organisms, pathogens, and insects. Nowadays, the stability of food supply is largely dependent on agrochemicals. Therefore, agrochemicals have to be harmless not only to higher plants and animals, but also to environmental ecosystems.

We typically call agrochemicals used to control disease "fungicides". This term came from the idea that killing plant pathogens is the best way to protect plants from diseases. However, the toxicants to microorganisms are considered to be more or less harmful to other organisms because the unit of life, the cell, is almost the same from the structural and functional viewpoints between higher organisms and microorganisms. Further, fungicides may kill the nonpathogenic microorganisms, saprophytes, which play a very important role in the recycling of organic matter produced annually. Thus, there are some risks in the use of toxicants because they cause environmental pollution.

Then what should we do? The answer is to develop disease control agents without toxicity, that is, compounds which do not kill microorganisms but inactivate the pathogenicity of pathogens. These may be the ideal control agents. If we could succeed in accomplishing this, theoretically we could solve the problems of environmental pollution. This would be possible because pathogenicity is a special property of plant pathogens and the compounds which specifically suppress pathogenicity do not injure any other organisms.

2. Effect of Nonpathogenic Microorganisms on Plant Leaf Surfaces (Epiphytic Microorganisms)

Many nonpathogenic microorganisms live on the healthy leaf surfaces of plants. They are classified into two groups, residential and casual microorganisms.[1] Residentials can grow and multiply on the surface of healthy, living parts of the shoot systems, and do not cause diseases. They derive nutrients from the exudate of plant surfaces. Casual microorganisms attach accidentally on plant surfaces. Several residential microorganisms have been found to compete with the pathogenic microorganism and reduce the outbreak of diseases.[2,3] For example, *Candida* spp. *Culvuraia* spp., *Cladosporium* spp., etc., which are isolated from the leaf surface of rice, reduce the incidence of *Helminthosporium* leaf spot disease when the spores of pathogenic fungus are inoculated with these fungi.[2] Many fungi and bacteria isolated from seeds and stems of peas are reported to compete with pea pathogens such as *Fusarium* and *Rhizoctonia*.[4] Residential microflora on corn kernels are protected from seedling blight, but sterilization of kernels with NaOCl nullifies the slight benefit conferred by residential organisms.[5]

These facts suggest that the application of "fungicides" causes a kind of resurgence later on, by killing competitors. The application of compounds that specifically suppress pathogenicity do not have such risks, because competitors may not be killed.

The purpose described in detail in Chapter 1 is to consider how to inactivate pathogenicity, in other words, to know the detailed nature of enemies in a plant protection program.

One point we should keep in mind is that compounds which specifically suppress pathogenicity are protectant of diseases but not eradicant or therapeutant and may not be effective if we apply these compounds after the establishment of infection by pathogens.

Theoretically, substances which have one or more of the following three properties have the possibility of being a disease control agent:

1. Kill the pathogen or inhibit the growth of the pathogen
2. Inactivate the pathogenicity without killing the pathogen
3. Enhance the resistant mechanisms of host plant

B. Fungicides

Though fungicides or fungistatic agents are not desirable for disease control agents from the point of view of environmental pollution, the use of these compounds is indispensable in cases where the infection of pathogens has already been established. Otherwise, pathogens grown on the infected host plants become the secondary inoculum to the neighboring healthy plants.

Fungicides that have been used in these cases are called eradicants and should have strong fungicidal activity with immediate effect. The residual period should be short. The selective toxicity is highly desirable for eradicants.

There are a lot of eradicants such as benomyl, carbendazim (MBC), dithiocarbamate fungicides, and so on. Among these, we will discuss briefly the mode of action of benomyl and of ergosterol biosynthesis inhibitors which contribute greatly to the knowledge of selective toxicity and the fundamental

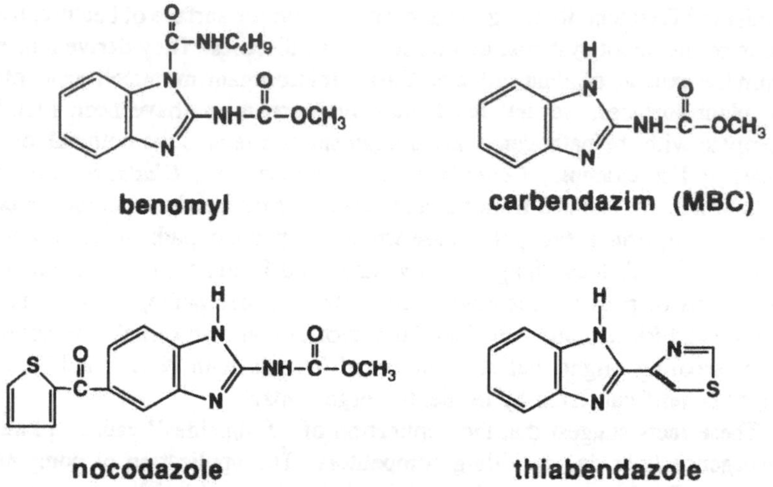

FIGURE 2. Benzimidazole fungicides.

sciences. As to the detailed description of benzimidazole fungicides, there is a review by Davidse.[6]

Previously it was believed that the mechanism of action of benzimidazole fungicides was the inhibition of nucleic acid synthesis of the fungi, but now it has been shown that these phenomena are the secondary effect arising from the inhibition of nuclear division. That is, benzimidazole fungicides such as benomyl, carbendazim (MBC), nocodazole, and thiobendazole (Figure 2) inhibit the formation of microtubles by binding the constitutive protein, tublin, and inhibit the formation of spindle fiber at nuclear division; and as a result, deoxyribonucleic acid (DNA) synthesis is inhibited. Though the microtubles are widely present in Eukaryotes, benzimidazole fungicides show a high selectivity to some fungal microtubles. For example, carbendazim is highly selective to fungi belonging to Ascomycetes, is lowly selective to Basidiomycetes, and shows no activity to Phycomycetes. Nocodazole is active against Oomycetes (Phycomycetes). Recent research clarified that these differences in activity of benzimidazoles against fungi are due to the difference of binding activity to tublins of each organism.

The appearance of tolerant fungi to benzimidazole fungicides is due to the decrease of binding activity of the fungicides to tublin, especially β-tublin.

Analyses of a full nucleotide sequence of genes encoding β-tublin from benzimidazole sensitive and tolerant *Neurospora crassa* indicate that the sensitivity to the fungicides depends on the 198th amino acid of β-tublin, glutamic acid. That is, the substitution of ^{198}glutamic acid to other amino acids by gene transformation decreases sensitivity. In the tolerant fungus, ^{198}glutamic acid is generally substituted by glycin. Diethofencarb (see Figure 3) is specifically toxic to many benzimidazole-resistant fungi having β-tublin of which the 198th amino acid is glycine, but not to the benzimidazole-sensitive fungi.

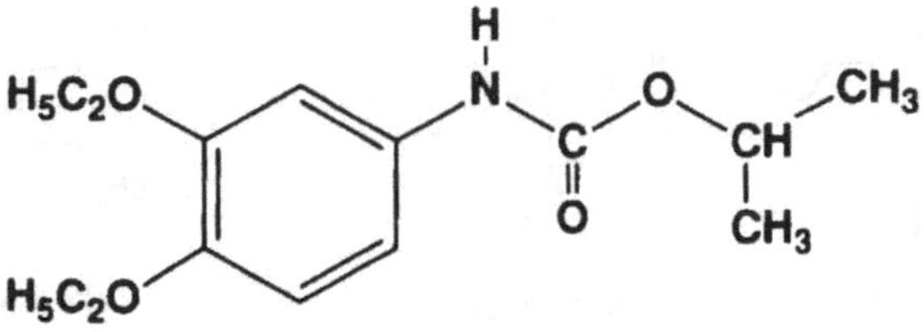

FIGURE 3. Diethofencarb.

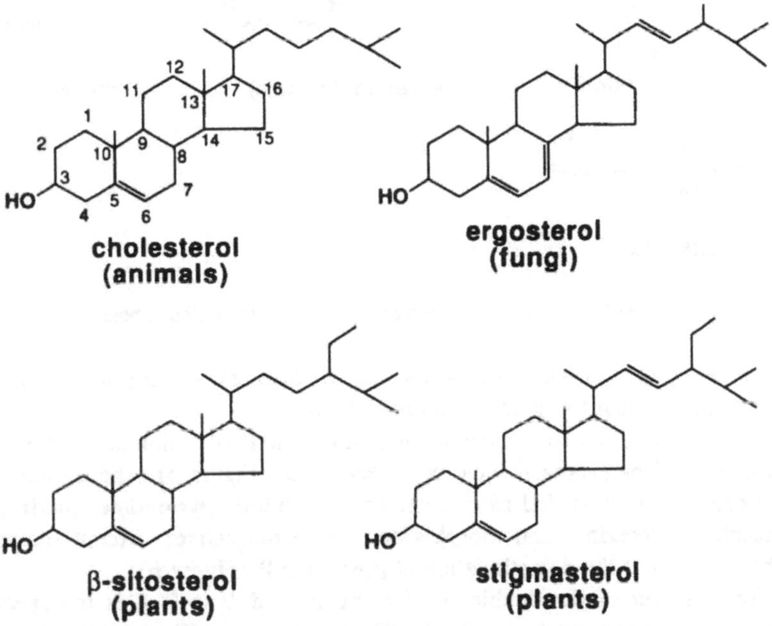

FIGURE 4. Sterol derivatives in living organisms.

The substitution of [198]glutamic acid or glycine with the other amino acid by gene transformation decreases the sensitivity to both fungicides.[7] The other most attractive fungicide for scientists is the ergosterol biosynthesis inhibitors.

Sterol derivatives distribute widely, and some of them are essential for living organisms. As indicated in Figure 4, cholesterol is the main sterol derivative in animals, β-sitosterol and stigmasterol are in the plant kingdom, and ergosterol is in fungi. These sterol derivatives have a perhydrophenanthrene ring, and are synthesized from acetate via the mevalonic acid pathway. These sterols differ from each other in the side chain at the 17 position or the position of double bond of B rings. Ergosterol is known to be present in the plasma membrane of fungi and plays an important role in the function of the membrane. Therefore, theoretically, specific inhibitors for ergosterol biosynthesis may be the specific toxicants to fungi; but practically, ergosterol biosynthesis inhibitors are more or less toxic to higher plants. According to

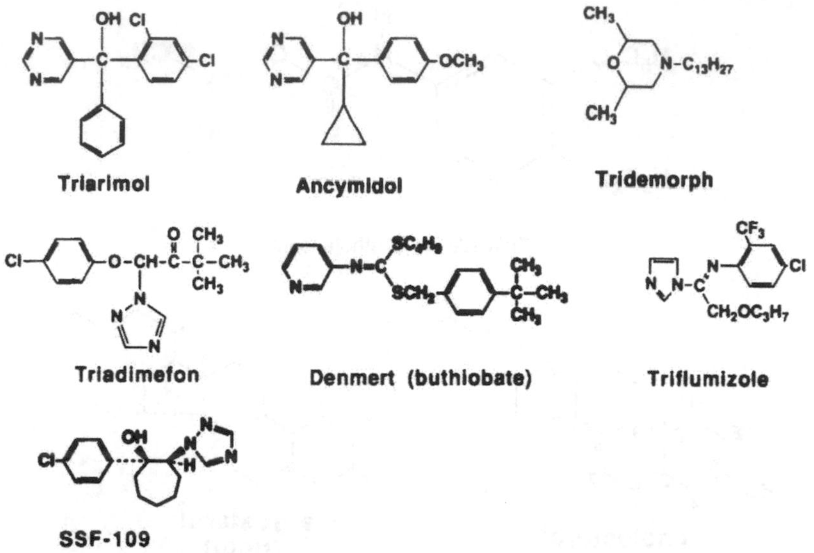

FIGURE 5. Fungicides which inhibit ergosterol biosynthesis.

Kraemer et al.,[8] compounds having the imidazol radical are more toxic to higher plants than the triazole compounds are.

At present, many fungicides for which the mode of action is by inhibitors of ergosterol biosynthesis have been developed (Figure 5). These inhibitors are chemically classified as derivatives of pyridine, pyrimidine, imidazole, triazole, pyperazine, and morphorine. These derivatives, except for morphorine, inhibit the demethylation at position 14[9] (Figure 6).

The transfer of the double bond at position 8–9 to 7–8 is the specific reaction in ergosterol biosynthesis. Therefore, a specific inhibitor for this process seems to be a selective toxicant to fungi without toxicity to higher organisms. Tridemorph acts in such a way.

These compounds are used to control powdery mildew, rust, smut disease, etc. of many crop plants. They are systemic fungicides and have a curative effect when used after the disease outbreak. Fungi treated with these compounds decrease in ergosterol content in the plasma membrane, and the transfer of substances in and out of cytoplasma is affected. Morphologically, the distortion of sporidia of rust fungi and haustoria of powdery mildew fungi occurs by the treatment with these fungicides, which results from the destruction of the membrane function.

II. COMPOUNDS WHICH INACTIVATE PATHOGENICITY

As described in Chapter 1, plant pathogenic fungi have a special property, pathogenicity, which is not present in saprophytic fungi. Therefore, com-

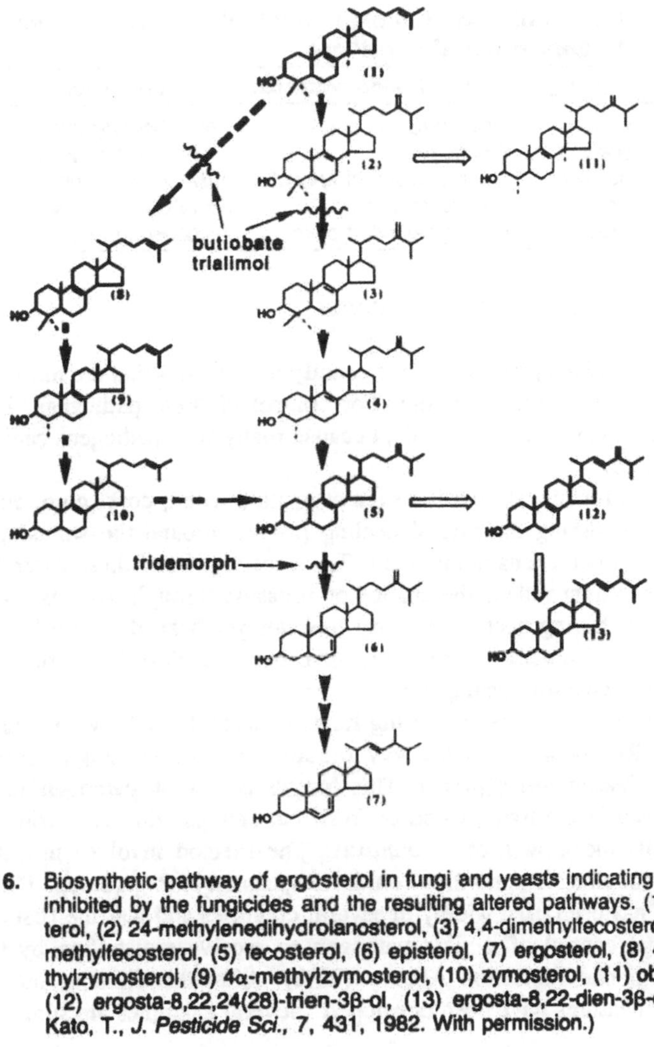

FIGURE 6. Biosynthetic pathway of ergosterol in fungi and yeasts indicating the sites inhibited by the fungicides and the resulting altered pathways. (1) Lanosterol, (2) 24-methylenedihydrolanosterol, (3) 4,4-dimethylfecosterol, (4) 4α-methylfecosterol, (5) fecosterol, (6) episterol, (7) ergosterol, (8) 4,4-dimethylzymosterol, (9) 4α-methylzymosterol, (10) zymosterol, (11) obtusifoliol, (12) ergosta-8,22,24(28)-trien-3β-ol, (13) ergosta-8,22-dien-3β-ol. (From Kato, T., *J. Pesticide Sci.*, 7, 431, 1982. With permission.)

pounds which inactivate some processes of pathogens to express pathogenicity might be ideal disease control agents that do not injure higher organisms and also useful microorganisms in the natural ecosystem.

A. Inhibition of Invasion Into Plants

The first requisite for plant pathogens is to invade into plant tissues or cells in order to derive nutrients from the plant. Therefore, if we could inhibit the ability of pathogens to enter into plant tissues or cells, the pathogens could not derive nutrients from plants and hence could not cause diseases.

Table 1. Control of Wood-Rotting Fungi Which Invade Through Wounds by Saprophytic Competitors

Wood-rotting fungi	Kinds of wood protected	Competitors	Ref.
Fomes annosus	Pine stump	*Peniophora gigantea*	10
Hymenomycetes	Red maple	*Trichoderma harzianum*	11, 12
Eutypa armeniacae	Pruning wound of apricot	*Fusarium lateritium*	13
Poria carbonica	Douglas fir pole	*Scytalidium lignicola*	14
Stereum purpureum	Pruning wound of plum	*Trichoderma viride*	15

1. Wound Infection and the Control

Lowly evolved plant pathogens usually use wounds that are made naturally or artificially as the entry site. For control of these pathogens, it is most important to avoid such wounds, because many such pathogens cause serious crop losses.

Barriers for fungal invasion such as wound wood, cork, gum, and tylosis are formed during the natural healing process around the wounds, and the invasion of pathogens is inhibited. The acceleration of these natural healing processes might reduce the chance of invasion through wounds, although a compound having such a property has not yet been developed. However, several useful biocontrol measures have been developed to control diseases caused by wound-infecting fungi.

Fomes annosus, a wood-rotting fungi invades through the cut end of pine stumps. Rishibeth[10] devised a way to control the rot by using an antagonistic fungus, *Peniophora gigantea*. This fungus is a weak pathogen and can be established in the living wood of fresh cut stumps without causing disease, but limits the growth of *F. annosus*. The method involves inoculation of freshly cut pine stumps with a conidia suspension of *P. gigantea*. This control method has been used widely in several countries and was the first agent for biological control of a plant pathogen to receive registration by the U.S. Environmental Protection Agency (EPA). After this finding, many useful control measures have been developed. Several examples are summarized in Table 1.

The crown gall of many plants caused by *Agrobacterium tumefaciens* enters through wounds. The biological control measure by inoculation of plant roots with a bacteriocin-producing bacterium, *Agrobacterium radiobacter*, has been widely used in many countries. This method is reviewed by Moore.[16]

2. Inhibition of Invasion Through Natural Openings

It seems reasonable to consider that this inhibition of stomatal entry of fungi by compounds which close stomata (such as phenylmercury acetate, chlorogenic acid, hydroxamine, etc.) might be an effective way to control diseases caused by such fungi. However, practically, the use of this method is likely to be ineffective, because downy mildew fungi and appressoria of uredospores of many rust fungi formed on stomata force entry into the substomatal cavity through closed stomata as described in Chapter 1.

3. Inhibition of Direct Penetration (Cuticular Penetration)

Many pathogenic fungi that cause serious diseases in important crops enter into plants by way of direct penetration (cuticular penetration) through intact plant surfaces. The blast fungus of rice, one of the most devastating plant pathogens in Asian countries, enters into the host plant by this way.

For the control of this disease, organic mercurial compounds had been used in Japan, but use was prohibited from the hygienic viewpoint. Since then, chemical companies in Japan have searched for a safe blast control agent as an alternative to mercurial compounds. As a result, pentachlorobenzylalcohol (PCBA) was found to be very effective in controlling the rice blast disease when it was applied before the fungal infection. This compound did not inhibit spore germination and mycelial growth of the blast fungus even at 1000 ppm. Therefore, the application of PCBA after the infection was completely ineffective. Spraying of this compound at 40 ppm on rice seedlings before the inoculation inhibited the disease incidence nearly completely.

To elucidate the mode of action of PCBA, Oku and Sumi[17] devised a model experimental system using an artificial membrane. That is, a cellophane sheet floating on the water extract of rice straw is perforated abundantly by the germ tubes of blast fungus. However, it was found that such a perforation could be inhibited completely when PCBA at a final concentration of 50 ppm was added to the rice straw extract. The germ tubes of *Helminthosporium* leaf blight fungus or the black rot pathogen of citrus, *Alternaria citri*, perforated the cellophane sheet even in the presence of PCBA. That is, the perforation-inhibiting activity of PCBA coincides completely with the specificity of PCBA to control diseases in the field. However, based on only the above experimental results, it is possible that some penetration-inducing factors might be present in the rice straw extract and that PCBA inactivates these factors. This possibility was denied because the perforation-inhibiting ability of PCBA was not reduced when the concentration of rice straw extract was increased. Further, PCBA did not inhibit cellulase activity produced by the blast fungus. Thus, the effect of PCBA in controlling the rice blast disease was found to inhibit the mechanical perforation ability of the pathogen into the rice plant.

Unfortuantely, PCBA was withdrawn from the market after a few years. The reason was that PCBA residing on rice straw was metabolized into chlorinated benzoic acid by some soil microorganisms when the PCBA-treated straw was used as the compost and caused malformation on vegetables which were grown on the compost.

However, since then active research on this type of "antipenetrant" has been conducted; and at present, many compounds of this type are on the market, and make up a large share of the control agents used for the rice

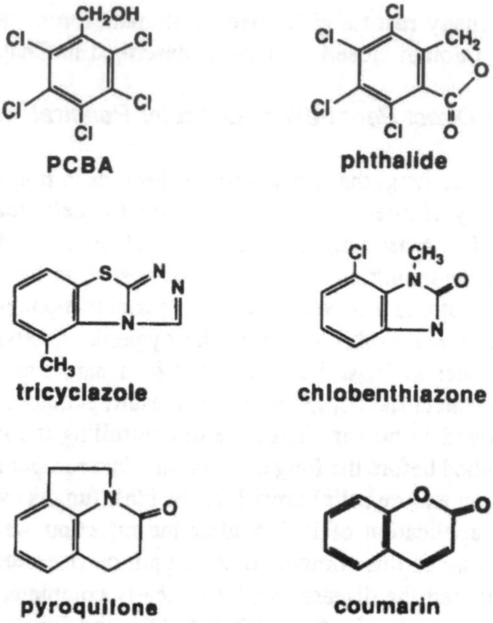

FIGURE 7. Control agents for rice blast disease acting as an antipenetrant for the pathogen.

blast disease. The main antipenetrants for the rice blast fungus are indicated in Figure 7.

On rice leaves sprayed with these compounds — such as 4,5,6,7-tetradichlorophthalide (Futharide), 1,2,5,6-tetrahydro 4H-pyrrolo-(3,2,1-ij) quinolin-4-one (Pyroquilone), 5-methyl-1,2,4-triazolo (3,4,6)-benzothiazole (Tricyclazole), etc. — conidiospores of the rice blast fungus germinate as usual and form abundant appressoria, but cannot penetrate into rice leaves. According to the careful observation of appressoria formed on rice leaves which are treated with such an antipenetrant, the melanization of appressoria is inhibited. Therefore, the search for melanization inhibitors is now one of the main projects for chemical industries. The importance of melanization of appressoria in penetration ability has been elucidated.[18–20]

Melanin, brown pigments distributed widely in plants, animals, microorganisms, and biosynthetic pathways in animals, has been studied extensively. The precursors and biosynthetic pathways of melanins are different in organisms, and also in microorganisms.

As shown in Figure 8, melanin in animals is synthesized from tyrosine via dioxyphenylalanine (DOPA), and is called DOPA melanin. In some molds, melanin is synthesized from shikimic acid via γ-glutamyl-4-hydroxybenzene (GDHB) and is called GDHB melanin. In other fungi the precursor of melanin is catechol, and the oxidation product, quinone, is polymerized to a brown

DOPA melanin

Tyrosine → DOPA → Melanin (many fungi, animals)

Catechol melanin

Catechol → Quinone → Melanin (*Ustilago maydis*)

GDHB melanin

Shikimic acid → GDHB(γ-glutaminyl-3,4-dihydroxybenzene) → Melanin (fruiting body of higher fungi)

DHN melanin

Acetate → Scytalone → Vermelone → DHN(dihydroxynaphthalene) → Melanin (*Verticillium* , *Pyricularia* , *Colletoricum* etc.)

FIGURE 8. Kinds of melanin in biological kingdom.

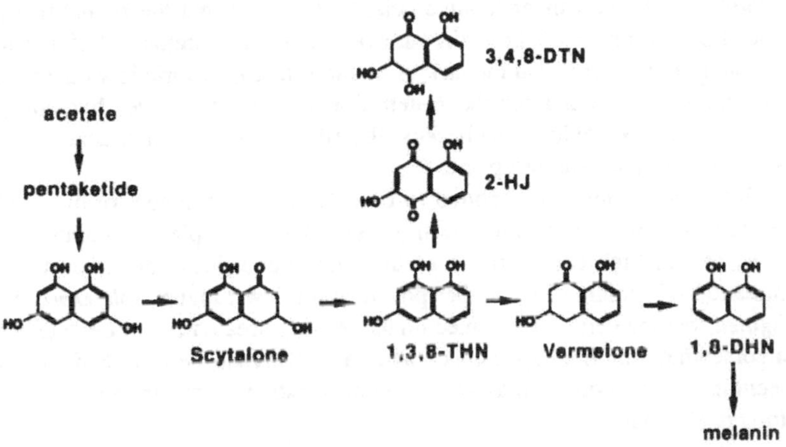

FIGURE 9. Rough scheme for biosynthetic pathway of DHN melanine. 1,3,8-THN: 1,3,8-trihydroxynaphthalene; 1,8-DHN: 1,8-dihydroxynaphthalene; 2-HJ: 2-hydroxyjuglone; 3,4,8-DTN: 3,4-dihydro-3,4,8-trihydroxy-1(2H)-naphthalene. (From Yamaguchi, I., *J. Pesticide Sci.*, 7, 314, 1982. With permission.)

pigment. This type of melanin is called catechol melanin. 1,8-Dihydroxynaphthalene (DHN) melanin, which is synthesized from acetate through penta ketide and DHN, was found in research using albino mutants of *Verticillium dahliae*. These melanins, in general, protect living organisms from environmental injury such as ultraviolet rays. The ink ejected by cuttlefish and octopuses also is melanin and protects these marine organisms from natural enemies.

The melanin in appressoria of several plant pathogenic fungi such as the rice blast fungus, *Verticillium* spp., anthracnose pathogens, etc. belongs to DHN melanin for which a rough scheme for biosynthesis is indicated in Figure 9; and the melanin was found to play an important role in the penetration ability of these pathogens.

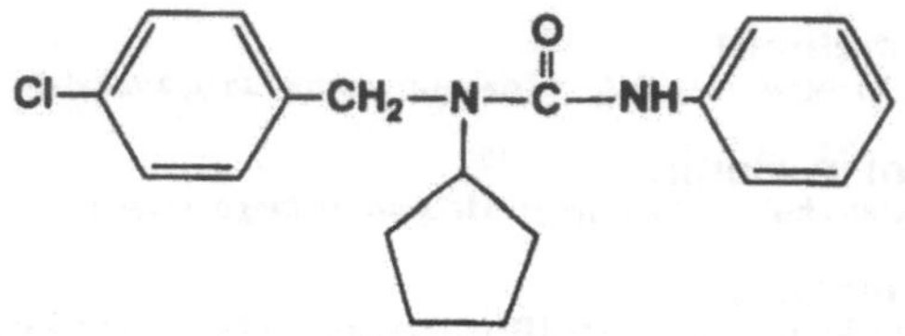

FIGURE 10. Pencyclone effective against rice sheath blight.

Antipenetrants indicated in Figure 7 have been known to inhibit between 1,3,8,-trihydroxynaphthalene (THN) and vermelone in the biosynthetic pathway.[20]

As to the mode of action of melanin biosynthesis inhibitors as disease control agents, two theories are available.[20] One is that the by-product produced by inhibition of the main pathway, 2-hydroxyjuglone (2-HJ) is toxic to fungi. The other is that the lack of melanin in appressoria loses the rigidity of appressoria needed for the penetration ability. The latter hypothesis is considered to be predominant because the albino mutant of anthracnose fungus also cannot penetrate into plants.[21]

Soybean lecithin was reported to be effective for the control of powdery mildew disease of vegetables such as cucumbers, eggplant, pumpkins, peppers, etc. A high concentration of this compound does not inhibit the germination of conidia of cucumber powdery mildew fungus, *Sphaerotheca fuliginea;* but the germ tube formed on the lecithin-treated cucumber leaf shows a somewhat abnormal appearance and cannot invade into the leaf. Soybean lecithin was also reported to inhibit the penetration of rice leaves by the blast fungus at 5 ppm.[22]

4. *Inhibition of Infection Cushion Formation*

The sheath blight of rice plants caused by *Rhizoctonia solani* is also a devastating disease in Asian countries because the infection of the sheath at the head-bearing stage severely inhibits the growth of ears and hence, rice grain production. This fungus invades the rice sheath by hyphae after forming a mycelial mat called the infection cushion.

One of the most important working mechanisms of validamycin A, an antibiotic,[23] is reported to inhibit the formation of an infection cushion of the rice sheath blight fungus; as a result the fungus cannot penetrate into the host plant.

Pencyclone (Figure 10), recently developed in Japan, is also specifically effective against the rice sheath blight and other diseases caused by *Rhizoctonia solani*. Although this compound is known to inhibit the mycelial growth of this fungus in vitro, the main mechanism of action in disease control is due to the inhibition of penetration ability of this fungus.[24]

An avirulent, *Rhizoctonia solani*-like fungus is isolated in high frequency from roots of gramineous, leguminous, and flax plants. This fungus is called BNR. A new biocontrol method to control root rot of turf grass and bean plants caused by *R. solani* has been devised by using this avirulent isolate.[25] BNR does not compete with virulent *Rhizoctonia* in vitro. Because BNR grows well on the surface of roots and hypocotyls of the bean plant, the mode of action of this fungus to control the virulent fungus seems to be due to the acquired resistance or competition of the infection sites. In general, formation of an infection cushion of *R. solani* is accelerated by the exudate from the root. The inoculation with BNR inhibits the formation. Because formation of the infection cushion of a virulent fungus does not recover after the elimination of BNR, the effect of BNR to control disease is considered to be due to the change of components of root exudate according to the metabolic change of seedlings by the parasitic activity of BNR.

The inhibition of penetration ability of pathogens into host plants is the inactivation of the first requisite for pathogenic fungi. This means that the protective effect of the natural host epidermis is fully exhibited.

B. Inactivation of the Ability of Pathogens to Overcome the Resistance of Hosts

As described, living plants protect themselves by versatile defense mechanisms to defend from the attack by microorganisms. Pathogens seem to have acquired the ability to overcome these defense mechanisms of their own host or hosts during the historical coevolutional process. Among these abilities of pathogens, suppressors play a very important role in some plant pathogenic fungi. Therefore, compounds which inhibit biosynthesis of the suppressor or inactivate activity of the suppressor seem to be useful as disease control agents. By inactivation of suppressor activity, the host defense reactions are fully expressed and invasion of pathogens may be inhibited; thus this type of disease control has a kind of biological control as the principle.

Research on suppressors has been taking place only about two decades; therefore there is no such compound of the above type on the market as a disease control agent. However, it seems to be a promising research field for agrochemicals, although it may be possible that one compound is necessary for each plant disease because the suppressor from the pathogen has host specificity and differs in each pathogen. At the present time, several suppressors are identified as glycopeptides; therefore, a substance which inhibits biosynthesis of glycopeptide suppressors in general may be a possible potent disease control agent.

C. Inactivation of the Ability of Pathogens to Evoke Disease

The third property of plant pathogens is the ability to evoke disease. The main mechanism of this ability is due to the production of toxins or deleterious enzymes. Therefore, compounds which inhibit the biosynthesis or inactivate the activities of these virulence factors may be useful as disease control agents.

1. Inhibition of Biosynthesis of Toxins

Many plant pathogenic fungi produce toxic metabolites (toxins). We sometimes find that the amount of toxin produced by pathogens into culture media is controlled by the quality and quantity of the components of culture media. These facts suggest that the productivity of toxins can be controlled chemically. Scientists usually wish to produce and isolate a large amount of toxins for the research on toxins. Inversely, for the control of plant diseases, scientists have to devise a way to inhibit the toxin production.

Lycomarasmin is a toxic metabolite produced by a melon pathogen, *Fusarium oxysporum* f. sp. *melonis*, into culture media. The importance of lycomarasmin as the virulence factor is not established, but production in a culture filtrate was reported to be inhibited completely by a terpenic acid derivative contained in the flower of *Helichrysmus arenarium*.[26] Productivity of fusaric acid, a toxin produced by many species of *Fusarium*, is known to be controlled by Zn^{2+} and the kinds of nitrogen sources added into culture media.[18,27,29]

As described in Chapter 1, a host-specific toxin is a primary determinant of pathogenicity for producer pathogens. Therefore, compounds specifically inhibiting the biosynthesis of the toxin might be disease control agents.

This type of agrochemical is not yet marketed, but several interesting studies have been conducted. That is, Nishimura and co-workers found that the biosynthesis of AK toxin, a host-specific toxin produced by the black spot pathogen of Japanese pear (*Alternaria alternata* Japanese pear pathotype), was inhibited by the antibiotic cerulenin or the amino acid methionine.[30] Tusge et al.[31] further demonstrated that spores of *A. alternata* Japanese pear pathotype treated with 20 ppm of cerulenin or 100 ppm of methionine lost pathogenicity almost completely as well as productivity of AK toxin during germination. The germination of spores, elongation of germ tubes, and formation of appressoria and penetration hyphae are not significantly inhibited. Cerulenin and methionine are also effective in reducing pathogenicity and production of host-specific toxins of the apple and strawberry pathotypes of *A. alternata*.

Unfortunately, these compounds are not used as disease control agents because cerulenin is toxic to vertebrates, and the pear fruits treated with methionine have a bad odor. However, this kind of investigation seems to be an important field for the development of promising control agents in the future.

2. Inactivation of Toxins

Compounds which inactivate the toxicity of toxins produced by pathogens may be useful as disease control agents. The toxicity of fusaric acid, a toxin

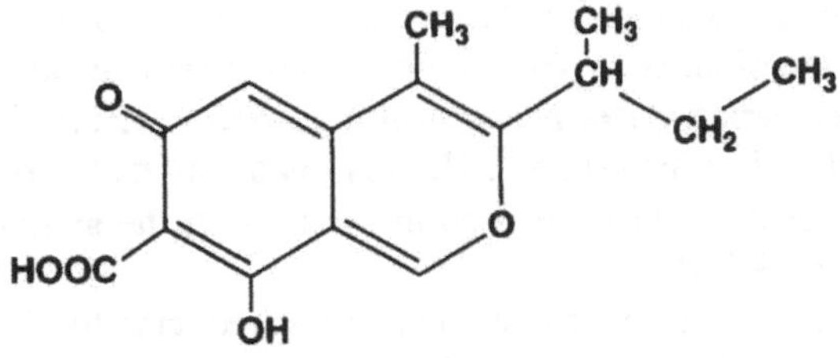

FIGURE 11. Ascochitine produced by *Ascochyta fabae*.

$$\text{glutamic acid} + NH_3 + ATP \underset{\textit{glutamine synthase}}{\overset{Mg}{\rightleftharpoons}} \text{glutamine} + ADP + Pi$$

FIGURE 12. Site of action of wildfire toxin, glutamine synthase.

produced by many species of *Fusarium*, was known to inactivate respiratory enzymes containing metals by forming chelate.[32-36] The inhibitory activity could be recovered by the addition of metal ions such as Fe, Co, Ni, Zn, Mn, and so on. This is due to the formation of metal chelate with fusaric acid and releasing the respiratory enzymes from fusaric acid. The toxins marticin and isomarticin produced by pea pathogen *Fusarium oxysporum* f. sp. *martii* are also inactivated by metal ions such as Cu, Al, and Fe. The addition of these metal ions into culture media reduces pathogenicity of the fungus when cultured on these media.[37]

Ascochitine (Figure 11), a metabolite produced by a broad bean pathogen, *Ascochyta fabae*, is phytotoxic and antibiotic. The phytotoxicity is inactivated by an equimolar concentration of Fe^{3+}.[38] The antibiotic activity of ascochitine is found to be inactivated by protein, alanine, asparagine, glutamine, and arginine.[39]

The toxicity of toxins can also be affected by environmental conditions such as pH, oxidoreduction potential, temperature, etc. For example, the minimum toxic dose of isomarticin to the pea plant is 8.5 mg/kg at pH 5.4, but is more than three times this at pH 7.5.[40] $Ca(OH)_2$ has always been known to affect the disease incidence, and it seems probable that the increase in pH value reduced the toxicity of toxins produced by pathogens.

Where toxins inhibit some sites of a metabolic pathway essential to maintaining plant health, the deficiency of metabolic intermediates behind the affected points causes trouble in the plant cells. In such cases, the artificial administration of deficient intermediates is sometimes effective in eliminating the toxicity. For example, wildfire toxin (tabtoxin) produced by *Pseudomonas syringae* pv. *tabaci* inhibits glutamine synthase as indicated in Figure 12; and as a result, glutamine is deficient in the affected plant cells. Therefore, the toxicity can be eliminated by addition of a high concentration of glutamine to the affected tobacco leaf.[41] Phaseolotoxin which is produced by the halo blight bacterium of the bean (*Pseudomonas syringae* pv. *phaseolicola*) inhibits ornithine transcarbamylase; hence ornithin accumulates, and citruline and arginine are deficient in the infected leaves. By addition of citruline or arginine to the center of the halo formed on the bean leaf, the symptom (halo) disappears within 48 hr.[42]

These examples of the inactivation of a toxin reducing the disease incidence are effective at the laboratory level but have not yet succeeded in controlling diseases of this type in the field. In practice, there seem to be difficulties in controlling disease by detoxification of toxins produced by pathogens, because if we could inactivate toxin for some period, the new toxin might be produced if the pathogenic microorganism were alive on the infected plant.

There is a trial to control the wilting disease of the tomato (caused by *F. oxysporum* f. sp. *lycopersici*) by biological degradation of fusaric acid with a microorganism. Hashimoto et al.[43] treated an avirulent strain of *Pseudomonas solanacearum*, originally a bacterial wilt pathogen of Solanaceae, with N-methyl-N′-nitro-N-nitrosoguanidine, and obtained a mutant tolerant to fusaric acid. They named the mutant strain A 16. The growth of A 16 was not inhibited at all in the medium containing 100 ppm of fusaric acid, and the toxin in the culture filtrate was completely detoxified after the cultivation. Inoculation with A 16 of a tomato cutting caused multiplication of A 16 within the cutting, and the inoculated cutting did not wilt by treatment with fusaric acid. In a greenhouse experiment, the *Fusarium* wilt of the tomato can be controlled by inoculation of the root with A 16 before the challenge with *F. oxysporum* f. sp. *lycopersici*. A 16 does not compete with *F. oxysporum lycopersici*.

In the meantime, there have been two theories on the mechanism for wilting of the tomato plant by *Fusarium*. One is the plugging theory of the vascular vessel with tyrosis or degrading products of vessel components by hydrolytic enzymes produced by the pathogen. The other is the toxin theory, by which the toxin, fusaric acid, might affect the plasma membrane of the host cell and enhance the permeability. The above results that the detoxification of fusaric acid eliminates the pathogenicity of toxin-producing fungi shows, though indirectly, that the toxin theory is predominant.

Recently genetic engineering has made progress in developing very effective control measures for plant diseases.

Anzai et al.[44] introduced the gene for detoxification of tabtoxin produced by *Pseudomonas syringae* pv. *tabaci* into the tobacco plant. The transgenic plant not only is tolerant to tabtoxin but also is resistant to this pathogen. This will be discussed later in Chapter 6, Section V. This experimental result suggests several important points. One is the new technology for the breeding of disease-resistant plants. The other shows the importance of tabtoxin as the virulence factor in *P. tabaci*, and by inactivation of the virulence factor plant diseases can be controlled.

III. INCREASE IN RESISTANCE OF CULTIVATED PLANTS

Other than the compounds which inhibit the function of plant pathogenic microorganisms, plant diseases can be controlled by compounds which enhance the resistance of plants.

A. Increase in Static Resistance

As described in Chapter 2, the static resistance is due to constitutional properties of the plant such as physical structures or chemical constituents.

Both the rice blast fungus and the *Helminthosporium* leaf blight fungus enter into rice leaves through motor cells. This entry site is used because the surface layer of the motor cell is less resistant to penetration than the other surface layer of the epidermal cells. It is well-known that the application of sodium silicate to the underground part of rice plants increases the resistance against these pathogens. Akai[45] concluded that the silicate specifically deposited on the surface layer of the motor cell and strengthened the physical resistance to penetration by these fungi. He observed that a larger amount of silicate accumulated at the position of motor cells of treated plants than that of untreated plants when both samples were mildly burned and the remaining ash was examined under a microscope. Ishiyama and Sato[46] examined the effect of alkylisocyanate derivatives (R-SCN) on the control activity of the rice blast disease and found that $C_{16}H_{33}SCN$ and $C_{18}H_{37}SCN$ were most effective. These compounds are effective when applied to rice leaves before inoculation, but not after invasion of the pathogen already had been established. Elimination of the cuticular layer, or elimination of wax on the leaf surface with ether before the treatment with these compounds eradicates the effect on disease control. However, the washing of leaves with water 1 hr after treatment with these compounds does not diminish the control effect. From these experimental results, they concluded that the mechanism of action of these compounds to control blast disease might be the increase in resistance of the epidermal wall to penetration by the blast fungus, forming a complex with wax.

Contents of many components of plants have been known to change by application of several agrochemicals. Particularly, many reports are available concerning the control agents for rice blast diseases. These works seem to be conducted to determine the side effects of fungicides, that is, whether the change of physiological conditions by fungicides affects the resistance of rice plants other than the direct fungicidal activity.

The spraying of O,O-diisopropyl S-benzyl phosphorothiolate (IBP) (Figure 13) and phenyl mercury acetate increases the content of total phenolics in rice leaves. The increase was reported to be especially greater in the latter. That is, the total phenolics increased 20–100% 1 day after spraying of phenyl mercury acetate. These results suggest that in addition to antifungal activity,

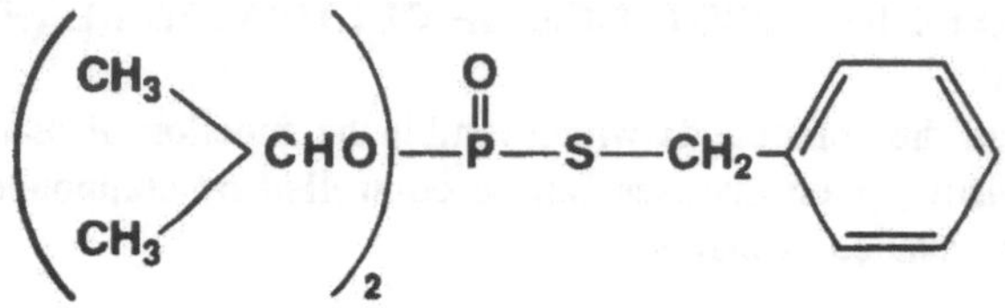

FIGURE 13. IBP effective against rice blast disease.

the blast control agents act to enhance resistance indirectly through the increment of antimicrobial components of the rice plant.[47]

B. Increase in Dynamic Resistance

The activities of the defense reaction of plants against pathogens are sometimes evaluated by estimation of phytoalexin productivity or hypersensitivity. It is known that the treatment of plants with some inorganic or organic compounds enhances the hypersensitivity and decreases the disease incidence.

According to Sempio,[48] cadmium nitrate acts as a systemic protectant against wheat powdery mildew disease. Mayer[49] examined the mechanism. That is, wheat grown on the water culture medium containing a low concentration of Cd became resistant to powdery mildew disease. On those wheat leaves, the conidia germinated, penetrated into the epidermal cells, and began to form haustoria 48 hr after inoculation; but the haustoria soon ceased to grow. The Cd content of treated wheat leaves did not reach the antifungal level, but hypersensitive reactions when infected were apparently enhanced.

Antibiotics applied systemically to a plant can increase the hypersensitivity. Müller et al.[50] found that streptomycin is effective against the late blight disease of potatoes caused by *Phytophthora infestans* to which this antibiotic does not show antifungal activity. Vörös et al.[51] reported that the treatment of potato tissue with streptomycin induces the increase of polyphenoloxidase activity, and concluded that the effect of streptomycin to the late blight disease may be an indirect one.

Several amino acids have been known to control the incidence of many plant diseases, and the outline is reviewed by Van Andel.[52] Application of about 1/100 *M* of methionine to the cut end of the stem of barley reduces the incidence of powdery mildew disease caused by *E. graminis* f. sp. *hordei*.[53] The order of effectiveness among stereoisomers is L-, DL-, and D-methionine. Because the spraying of L-methionine on mycelia formed on barley leaves does not inhibit the growth of the fungus, the mode of action of L-methionine is not due to antifungal activity. On the leaf surface which has absorbed methionine, conidia of the powdery mildew fungus normally germinate, form appressoria, penetrate, and begin to form haustoria; however, the abnormal haustoria arise and the absorption of nutrients may be suppressed markedly. The shape of haustoria formed in the epidermal cell was very similar to that

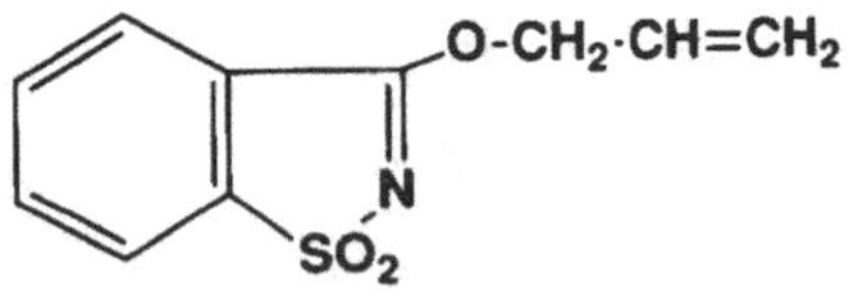

FIGURE 14. Probenazole effective against rice blast disease.

formed in the resistant cultivar. The electron microscopic observation[54] on the process of infection in L-methionine-treated barley showed that no degradation of haustorial cytoplasm occurs at the early stage of infection. At a later stage, the haustorial content increases in electron density; and large, more densely stained deposits are found in the cytoplasm. Epidermal cells of L-methionine-treated leaves become necrotic, accumulating electron-dense granules in vacuoles following inoculation. Because this phenomenon is observed in leaves of resistant varieties, it was concluded that the mode of action of L-methionine for control of powdery mildew might be enhancement of the resistance response of the host plant.

L-Methionine has the suppressive activity of disease incidence caused by fungi in which parasitism is highly advanced, but does not show the suppressive activity for diseases caused by fungi in which parasitic abilities are primitive, e.g., *Botrytis cinerea.*[53]

3-Allyroxy-1,2-benzisothiazole 1,1-dioxide, probenazole (Figure 14), a rice blast control agent is widely used in Japan. This compound has been developed as the systemic protectant when applied to paddy fields to diminish the labor of farmers.

Application of less than 200 g of this compound on the surface of 10 a of paddy field gives an excellent control effect toward the rice blast disease.[55] Probenazole and the metabolites in the rice plant show scarce antifungal activity; therefore it is hardly considered that the effect on rice blast disease is due to the antifungal action of this compound, because the concentration of this compound systemically distributed in rice leaves is very low. The main mechanism of action of probenazole is considered to increase resistance of the rice plant against the blast fungus. On rice leaves of which the root system was treated with probenazole, the penetration of the blast fungus is inhibited; and the invaded hyphae are suppressed to expand into neighboring host cells. Further, rice leaves treated with probenazole produce higher amounts of phytoalexin-like compounds than untreated ones when the blast fungus is inoculated. As for the phytoalexin-like compounds, Sekizawa and Watanabe[56] isolated three compounds; one, linolenic acid, accumulates 100–200 μg/g of treated leaves after inoculation with the blast fungus.

Dichlorocyclopropane derivatives are systemics and some of them are known to be effective as control agents of the rice blast disease. These derivatives show weak antifungal activity under certain conditions, but the high efficacy in controlling blast disease is hardly considered to be due to the antifungal

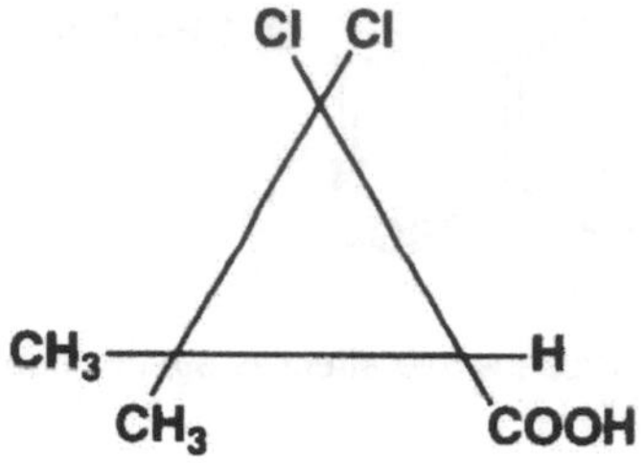

FIGURE 15. Dichlorocyclopropane carboxylic acid effective against rice blast disease.

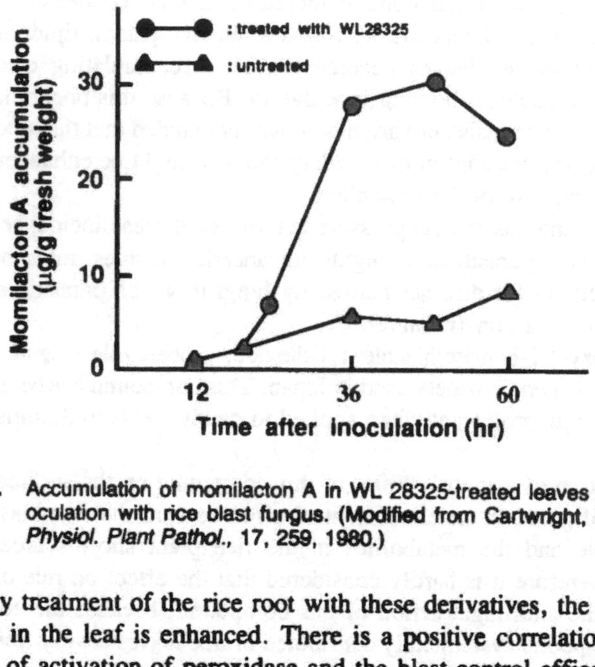

FIGURE 16. Accumulation of momilacton A in WL 28325-treated leaves following inoculation with rice blast fungus. (Modified from Cartwright, D. W. et al., *Physiol. Plant Pathol.*, 17, 259, 1980.)

activity. By treatment of the rice root with these derivatives, the activity of peroxidase in the leaf is enhanced. There is a positive correlation between the degree of activation of peroxidase and the blast control efficacy among these derivatives. Treatment of the rice root with the most effective control agent among these derivatives, 2,2-dichloro-3,3-dimethyl cyclopropane carboxylic acid (WL 28325, see Figure 15) enhances peroxidase activity of rice leaves about 12 times more than untreated ones. The inoculation of leaves of susceptible rice cultivars in which the root was treated with this compound shows the hypersensitive reaction-like browning, occurring similarly to inoculation of the resistant cultivars. The enhancement of peroxidase activity is considered to contribute the hypersensitivity-like reaction.[57] Further, in leaves in which the root was treated with WL 28325, momilacton A (a phytoalexin of rice), accumulates 10 times more as compared with untreated ones 48 hr after inoculation (Figure 16). Momilacton B does not accumulate in the untreated rice leaves that are inoculated, but accumulates at a concen-

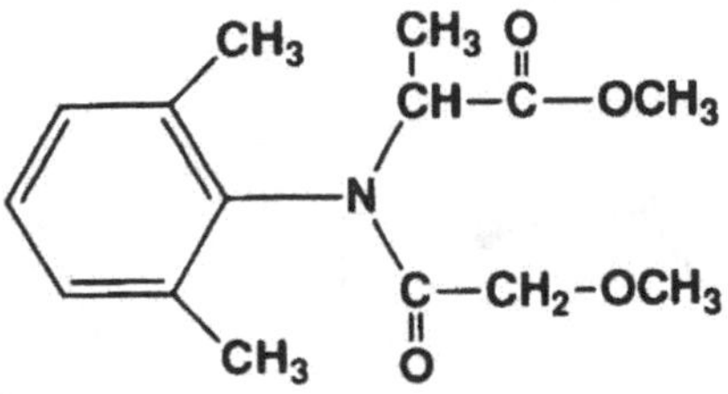

FIGURE 17. Metalaxyl effective against diseases caused by Phycomycetes.

tration of 7 μg/g in the treated rice leaves; and this is enough concentration to inhibit the growth of the blast fungus. From these results, the effect of WL 28325 is considered to be due to enhancement of defense activity of the rice plant against the blast fungus.[58]

Downy mildew and late blight fungus belonging to Oomycetes cause serious loss of crops and vegetables all over the world. To control these diseases, agrochemicals have been repeatedly applied. However, in the 1970s very useful systemic control agents were found, and they have contributed much to the simplification of control measures for these diseases. These compounds reside within the plant for a long time, hence reducing the frequency of application of agrochemicals. For example, dipping the pineapple root once in the solution of fosetyl-Al protects the pineapple from the rotting caused by *Phytophthora* for 18 months.

Methyl-N-(2-methoxyacetyl)-N-(2,6-xylyl)-DL-alaninate, metalaxyl (Figure 17), is an excellent systemic disease control agent for diseases caused by Phycomycetes, especially Peronosporales; it is widely used not only as a protectant of crops in the field but also as a control agent for postharvest diseases of fruits and tubers.

There are two theories on the working mechanism of metalaxyl. One is that the effect is due to enhancement of the resistance mechanism of host plants, and the other is that the enhancement of resistance is not a cause but a result of the antifungal activity.

According to Ward et al.,[59] the hypocotyl of soybean seedlings inoculated with compatible race 6 of *Phytophthora megasperma* f. sp. *glycinea* develops typical watersoak lesions; but when the roots of seedlings are treated with metalaxyl, the lesion development is suppressed. By treatment of roots with 20 μg/mL of metalaxyl, the hypocotyl lesions become restricted, brown, necrotic and indistinguishable from lesions formed in the untreated seedling inoculated with incompatible race 4. Further, a high concentration of the phytoalexin of soybeans, glyceollin, accumulates 12 hr after inoculation of metalaxyl-treated seedlings with compatible race 6 (Figure 18). This accumulation pattern is also similar to that in the incompatible interaction. To analyze the mechanism of action of metalaxyl from these results, Ward et al.[59] pointed out that glyceollin accumulated in leaves — possibly playing a critical role in preventing disease development — even if the concentration

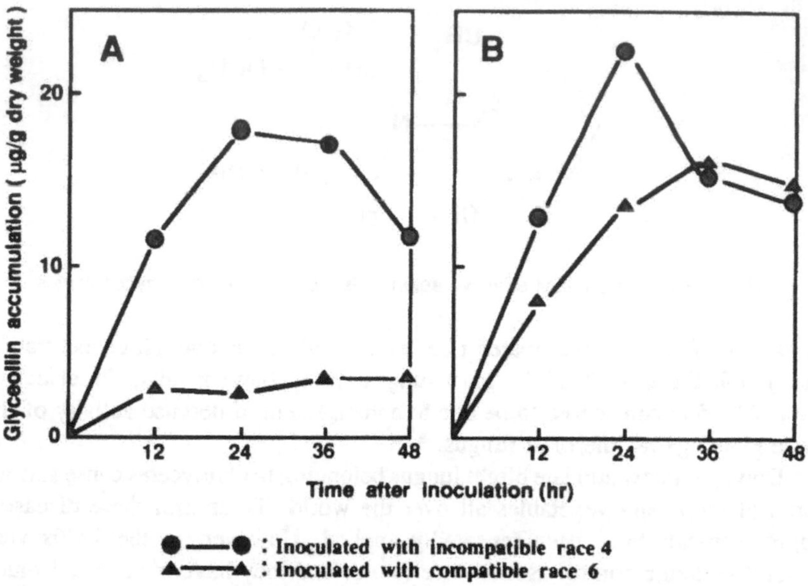

FIGURE 18. Effect of metalaxyl on glyceollin accumulation in soybean hypocotyls following inoculation with *Phytophthora megasperma* var. *sojae*. (A) Control (untreated); (B) root was treated with 20 μg/mL of metalaxyl. (Modified from Ward, E. W. B. et al., *Phytopathology*, 70, 738, 1980.)

of metalaxyl at the infection sites is less than fully inhibitory. However, the following possibility can not be ruled out: that the antifungal activity of metalaxyl affects the physiology of the fungus and as a result an elicitor-like substance is released from the affected fungus, just as pointed out by Király et al.[60]

Lazarovits and Ward[61] found that dipping the roots of soybean seedlings in a solution of 0.5 μg/mL of metalaxyl did not inhibit enlargement of lesions caused by the pathogenic fungus, in spite of the glyceollin content reaching an ED_{90} value for the fungus after inoculation. Treatment of the root at higher concentrations of metalaxyl restricts lesion enlargement. From these results, they consider that the protective effect is dependent on the antifungal activity of metalaxyl, and they question whether glyceollin plays a role in resistance.

Barak et al.,[62] on the other hand, reported that the spraying of potato leaf surfaces with metalaxyl before the harvest protects the postharvest rotting of tubers caused not only by late blight fungus but also by ***Fusarium sambusinum***, ***F. culmorum***, and ***Alternaria solani***. Because metalaxyl does not show any antifungal activity to the latter three fungal pathogens, they believe that the activation of resistance response to these pathogens may be the mechanism of action. They also obtained evidence that the treatment of potato tuber disks with a nonantifungal concentration of metalaxyl (0.003–0.03 μg per disk) rendered these disks to be resistant to ***P. infestans*** and also to ***F. sambusinum***.

In these treated disks, the polyphenoloxidase activity increased to two times that of the untreated disks 25 hr after metalaxyl treatment; hence, they consider that metalaxyl enhances the resistance of potato disks through activation of the lignin biosynthetic pathway.

Börner et al.[63] determined the effect of metalaxyl on the accumulation and distribution in tissues of glyceollin after inoculation with an incompatible and a compatible race of *Phytophthora megasperma* f. sp. *glycinea*. The concentration of glyceollin in the incompatible interaction reached a maximum at the infection site, but fell sharply toward the uninfected tissue. In contrast, glyceollin concentration in the infected compatible tissue was about one quarter of that in the incompatible interaction at the infection sites. However, the characteristic high and localized glyceollin accumulation which occurs in the resistant response is mimicked in the compatible interaction if the tissues are treated with metalaxyl prior to inoculation. Further, metalaxyl has no effect on the level of phenylalanine ammonia-lyase which is one of the key enzymes in glyceollin synthesis induced by infection with a compatible race, or on the extent of glyceollin accumulation elicited by treatment with a glucan elicitor from the pathogenic fungus. They concluded that metalaxyl does not increase the capacity of the seedling to synthesize glyceollin, but the response of the host (high accumulation of glyceollin at the infection sites) is altered by metalaxyl just as the incompatible interaction. They noticed the importance of analysis at the infection site to evaluate the role of phytoalexins. Further, as demonstrated by Moesta and Grisebach,[64] the inhibition of glyceollin biosynthesis by L-2-aminooxy-3-phenylpropionic acid removes the resistance of soybeans, and thus they do not agree with Ward's theory that glyceollin may not contribute to the resistance.

In opposition to these theories, Ward[65] conducted the following experiments to ascertain the negative role of glyceollin in disease resistance. That is, he tested the effect of glyphosate, an herbicide known to inhibit the glyceollin biosynthesis by blocking the shikimate pathway,[66] on the disease control effectiveness of metalaxyl. Theoretically, if glyphosate reduces the diseases control activity of metalaxyl, the activation of glyceollin biosynthesis induced by metalaxyl may play some role in the disease control. The result was that glyphosate reduced the ability of metalaxyl at marginally inhibitory concentrations to restrict spread of *P. megasperma* f. sp. *glycinea* and also glyceollin production in inoculated bean hypocotyls. No evidence was obtained showing that glyphosate suppresses the toxicity of metalaxyl in vitro and the uptake of metalaxyl by soybean hypocotyls. And thus Ward came to the conclusion that at marginally inhibitory concentrations the activation of host defense mechanisms contributes to the disease control efficacy of metalaxyl.

However, later on Cahill and Ward[67] produced many metalaxyl-tolerant mutants of the pathogenic fungus by UV irradiation and studied whether metalaxyl is effective to diseases caused by these tolerant isolates. When metalaxyl-tolerant isolates are inoculated on hypocotyls of compatible soybeans

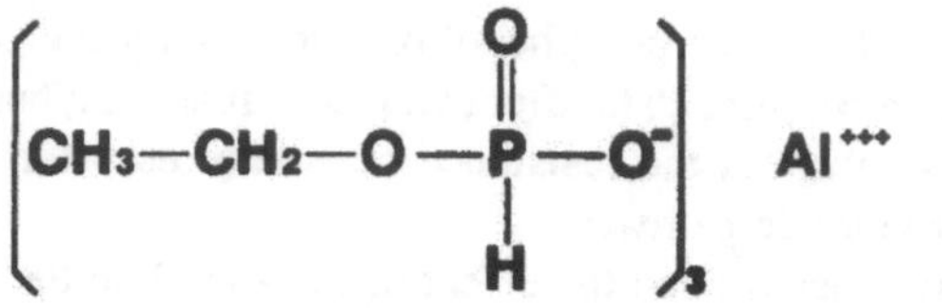

FIGURE 19. Fosetyl-Al effective against diseases caused by Phycomycetes.

treated with metalaxyl, a majority causes symptoms typical of compatible interactions with low levels of glyceollin. However, a few isolates that tolerate 500 μg/mL of metalaxyl in vitro are sensitive in inoculated plants and develop restricted lesions. As a result of comparing the release of the elicitor for glyceollin into culture fluids by metalaxyl-sensitive and tolerant isolates following addition of metalaxyl to the culture medium, they found that the elicitor is released by metalaxyl treatment in the culture medium of sensitive isolates but not in those of tolerant isolates. They concluded that in general these results are consistent with the possibility that in infected plants, at a marginally inhibitory concentration of metalaxyl, release of elicitors from fungal hyphae may stimulate the host defense responses in otherwise compatible interaction. However, further study is needed on the mechanism that metalaxyl-tolerant isolates in vitro are sensitive in vivo.

Thus, the mechanism of action of metalaxyl is very complicated; however, it is likely that this compound enhances the sensitivity of host plants to respond to pathogens, because the postharvest diseases of the potato tuber caused by fungi insensitive to metalaxyl such as *Fusarium* and *Alternaria* can be controlled by this compound. The antifungal activity of this compound in disease control activity cannot be ruled out because of the appearance of field isolates which cannot be controlled by this compound.

Aluminum salt of tris-*o*-ethyl phosphonate, fosetyl-Al (Figure 19), is also a commercially available systemic disease control agent effective against important diseases caused by Oomycetes. This systemic compound is known to move basipetalic direction within plants (symplastic movement) and hence may be useful in controlling root diseases by foliage application.[68]

Fosetyl-Al prevents growth of *Phytophthora* spp. in vitro only at concentrations of 1000 μg/mL or greater (EC_{50} 250 or more).[69,70] Therefore, it has been proposed that this compound acts indirectly by triggering a host-defense response. In contrast, several workers suggest that fosetyl-Al is metabolized in the plant tissue; and the breakdown product, phosphorous acid and phosphonic acid (H_3PO_3), may act directly against the pathogens by inhibiting mycelial growth, because the breakdown product is more fungitoxic.[71,72] The latter hypothesis is supported by the evidence that fosetyl-Al is ineffective to diseases caused by the tolerant strains of *Phytophthora* to phosphite (phosphorous acid ester).[73,74]

The basis of the hypothesis that fosetyl-Al enhances the resistance response of the host plant against pathogens is as follows. That is, when the fruits of

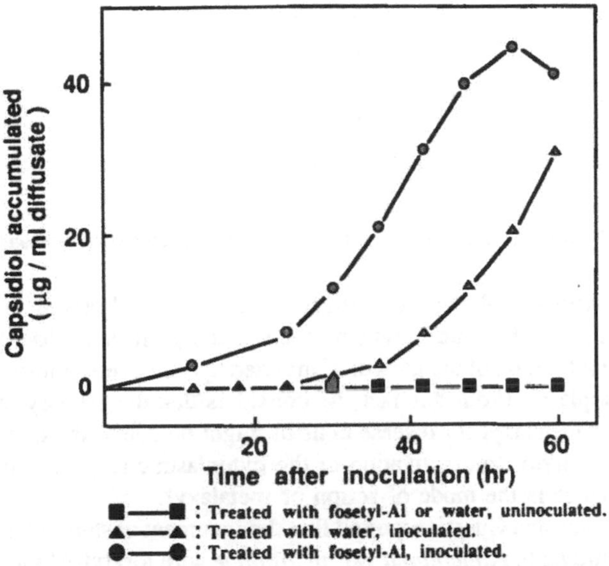

FIGURE 20. Effect of fosetyl-Al (100 ppm) on the accumulation of capsidiol in red pepper fruits following inoculation with *Phytophthora nicotianae* var. *parasitica*. (Modified from Guest, D. I., *Physiol. Plant Pathol.*, 25, 125, 1984.)

fosetyl-treated (100 µg/mL) pepper are inoculated with *Phytophthora nicotianae* f. sp. *parasitica*, a large amount of capsidiol, a phytoalexin of pepper, accumulates just as resistant cultivars are inoculated (indicated in Figure 20) and the lesions cease to grow.[75] The inoculation of susceptible tobacco in which the root had been treated with fosetyl-Al with *P. nicotianae* f. sp. *nicotianae* induces the accumulation of a large amount of capsidiol, and the strong hypersensitive reaction occurs around the lesions. Because fosetyl-Al inhibits the activity of elicitors produced by pathogens, the amount of capsidiol decreases when the fosetyl-treated resistant cultivar is inoculated. From these results, the mechanism of the disease control efficacy of fosetyl-Al is considered to be due to enhancement of the phytoalexin-producing ability of host plants.

Guest[76] further ascertained by a periodic microscopic observation of tobacco seedlings which had been treated with fosetyl-Al or phosphorous acid, that the inoculation with zoospores of *P. nicotianae* f. sp. *nicotianae* induces the same defense reactions that occur when the resistant cultivar is inoculated. That is, papilla is formed inside the cell wall of the infected cell, cytoplasm is aggregated very rapidly, nuclei of cells surrounding the infected cell migrate toward the site of infected cell, and then the infected hypha of the pathogen ceases to grow. The hypersensitive reaction has been observed ca. 20 hr after inoculation.

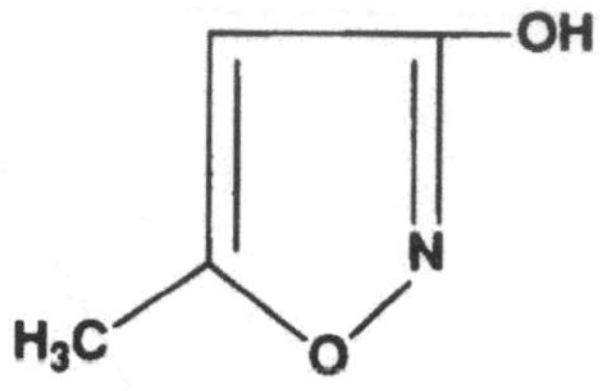

FIGURE 21. Hymexazole effective against soilborne plant diseases.

Along with the above experiment, Guest[76] also observed the infection process of the pathogenic fungus on the metalaxyl-treated tobacco cells, and found that inhibition of the growth of invaded hypha occurs before aggregation of the cytoplasm. From this fact, he considers that the primary factor of the efficacy of metalaxyl for disease control might be due to the direct action of antifungal activity, and activation of the cytoplasmic response appears to be a consequence in the mode of action of metalaxyl.

Nemestothy and Guest[77] reported that the treatment of stems of a susceptible tobacco cultivar to *P. nicotiana* var. *nicotianae* with fosetyl-Al induces a more rapid increase in sesquiterpenoid phytoalexins, lignin and ethylene accumulation, and phenylalanine ammonia-lyase activity after inoculation than in untreated stems. Propylene oxide and a phytotoxin, mevinolin, specific inhibitors of sesquiterpenoid biosynthesis; and aminooxyacetic acid, a nonspecific amino acid-transferase inhibitor, inactivate the effectiveness of fosetyl-Al in the susceptible tobacco cultivar. These results also support the theory that the mode of action of fosetyl-Al is enhancement of the host-defense response.

Abu-Jawdah and Kummert[78] reported that the foliar spray of bean leaves with fosetyl-Al delays the appearance of necrotic local lesions and reduces their number following inoculation with the alfalfa mosaic virus (AMV). Further, the isozyme pattern of the fosetyl-treated bean leaves changes after inoculation with AMV from that of untreated-inoculated ones. These facts can hardly be explained by the theory that antifungal activity is the main mode of action of fosetyl-Al.

3-Hydroxy-5-methylisoxazole, hymexazole (Figure 21), is a weak fungicide as indicated in Table 2, but is widely used in Japan to control the damping off disease of many plants caused by soil-inhabiting microorganisms including insensitive fungi such as *Fusarium* spp. Hymexazole is especially effective against the damping off disease of rice seedlings, cucumbers, watermelons, sugar beets, carnations, etc. which are caused by *Fusarium roseum*, *F. solani*, *Rhizoctonia solani*, *Pythium* spp., and so on.[79]

Concerning the mode of action of this compound, several in vitro data are available. That is, this compound inhibits the uptake of nutrients by inactivating the function of plasma membranes of pathogenic fungi,[80] or nucleic acid synthesis.[81] However, the concentration of hymexazole active to the fungi in laboratory experiments is much higher than the actual concentration

Table 2. Antimicrobial Activity of Hymexazole on Potato-Sucrose Agar

Pathogen	Minimum growth inhibitory concentration (μg/mL)
Alternaria kikuchiana	1000–300
Botrytis cinerea	300
Cercospora beticola	300
Cochliopolus miyabeanus	300–100
Corticum rolfisii	300–100
Corynebacterium michiganense	>1000
Fusarium oxysporum f. *lycopersici*	1000
F. oxysporum f. *melonis*	1000
F. oxysporum f. *niveum*	1000
Golmerella cingulata	100
Monilinia laxa	300
Mycosphaerella melonis	300
Pellicularia sasakii	1000
Phytophthora capsici	>1000
P. melonis	>1000
Pseudomonas solanacearum	>1000
Pyricularia oryzae	100–30
Pythium debaryanum	10–30
Rhizoctonia solani	300
Rosellinia necatrix	30
Sclerotinia sclerotiorum	300

Source: Tomita et al.[79]

which is used in the field to control the above diseases. For example, the minimum inhibitory concentration of hymexazole to *Fusarium oxysporum* is 1000 ppm in vitro as shown in Table 2; and the ED_{50} value for inhibition of deoxyribonucleic acid (DNA) synthesis of this fungus is 300 ppm.[81] However, the dose of this compound used to control *Fusarium* disease in the field is 3 L/m^2 of 500–1000 times diluent of 30% solution, namely, 1–2 g/m^2. Calculation of this dose in parts per million in soil gives 0.5 ppm and is sufficient in the field if we assume that 1–2 g of the active ingredient can control pathogens to a depth of 20 cm in the soil. Thus, it is hardly considered that the working mechanism presented by laboratory experiments acts as the control mechanism of these soilborne diseases in the field.

In practice, hymexazole is also used as the growth-promoting compound to grow healthy rice seedlings. Hymexazole is known to be metabolized within plant tissues into two kinds of glucosides, N-β-glucoside and O-β-glucoside (Figure 22).[82] Of these, hymexazole and N-β-glucoside exhibit promoting activity on root growth of rice seedlings. These facts suggest that the disease control efficacy of this compound is, in part, due to the increment of resistance of host plants and affects host physiology.

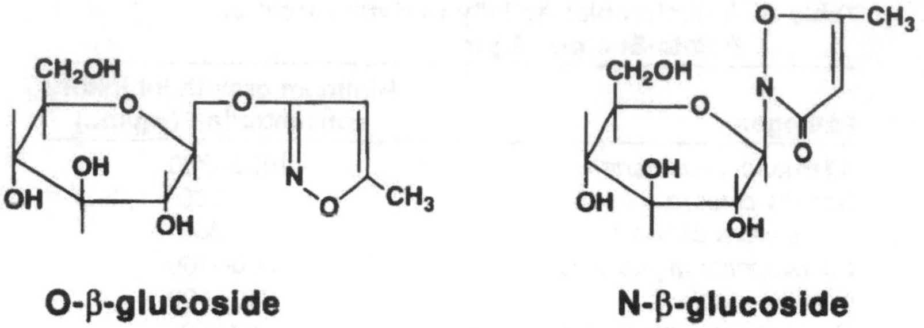

FIGURE 22. Two metabolites of hymexazole in plants.

REFERENCES

1. Leben, C., Epiphytic microorganisms in relation to plant disease, *Annu. Rev. Phytopathol.*, 3, 209, 1965.
2. Akai, S., Recent advance in studying the mechanism of fungal infection in plants, *Shokubutsu Byogai Kenkyu*, 8, 1, 1973.
3. Garret, S. D., Inoculum potential, in *Plant Pathology III*. Dimond, A. E. and Horsfall, J. G., Eds., Academic Press, New York, 1960, 23.
4. Kommedahl, T. and Windels, C. E., Evaluation of biological seed treatment for controlling root diseases of pea, *Phytopathology*, 68, 1087, 1978.
5. Lang, D. S. and Kommedahl, T., Factors affecting efficacy of *Bacillus subtilis* and other bacteria as corn seed treatment, *Proc. Am. Phytopathol. Soc.*, (Abstr.), 3, 272, 1976.
6. Davidse, L. C., Benzimidazole fungicides: mechanism of action and biological impact, *Annu. Rev. Phytopathol.*, 24, 43, 1986.
7. Fujimura, M., Kamakura, T., Oguri, Y., Inoue, H., and Yamaguchi, I., Analysis of the amino acid sequence of β-tublin in Dietofencarb and MBC sensitivity of *Neurospora crassa, Abstr. Pap. 17th Annu. Meet. Pest. Sci. Soc. Jpn.*, 1992, 44 (in Japanese).
8. Kraemer, W., Buechel, K., Meiser, W., Brandes, W., Kaspers, H., and Scheipflug, H., Triazol-O,N-acetals, chemistry, activity and structure, *Adv. Pest. Sci.*, 2, 274, 1979.
9. Kato, T., Mechanism of action of the fungicide buthiobate, *J. Pest. Sci.*, 7, 427, 1982.
10. Rishibeth, J., Stump protection against *Fomes annosus*. III. Inoculation with *Peniphora gigantea, Ann. Appl. Biol.*, 52, 63, 1963.
11. Pottle, H. W., Shigo, A. L., and Blanchard, R. O., Biological control of wound Hymenomycetes by *Trichoderma harzianum, Plant Dis. Rep.*, 61, 687, 1977.
12. Smith, K. T., Blanchard, R. O., and Shortle, W. C., Effect of spore load of *Trichoderma harzianum* on wood-invading fungi and volume of discolored wood associated wounds of *Acer rubrum, Plant Dis. Rep.*, 63, 1070, 1979.
13. Carter, M. V., Biological control of *Eutypa armeniacae, Aust. J. Exp. Agric. Anim. Husb.*, 11, 687, 1971.

14. Ricard, J. L. and Bollen, W. B., Inhibition of *Poria carbonica* by *Scytalidium* sp., an imperfect fungus isolated from Douglas-fir poles, *Can. J. Bot.*, 46, 643, 1968.
15. Grosclaude, C., Ricard, J., and Dubos, B., Inoculation of *Trichoderma viride* spores via pruning shears for biological control of *Stereum purpureum* on plum tree wounds, *Plant Dis. Rep.*, 57, 25, 1973.
16. Moore, L. W., *Agrobacterium radiobacter* strain 84 and biological control of crown gall, *Annu. Rev. Phytopathol.*, 17, 163, 1979.
17. Oku, H. and Sumi, H., Mode of action of pentachlorobenzyl alcohol, a rice blast control agent. Inhibition of hyphal penetration of *Pyricularia oryzae* through artificial membrane, *Ann. Phytopathol. Soc. Jpn.*, 34, 242, 1968.
18. Matsuura, K., Effect of melanin synthesis inhibitors on appressorial function in plant pathogenic fungi, *J. Pest. Sci.*, 8, 379, 1983.
19. Woloshuk, C. P. and Sisler, H. D., Tricyclazole, pyroquilon, tetrachlorophthalide, PCBA, coumarin and related compounds inhibit melanization and epidermal penetration by *Pyricularia oryzae, J. Pest. Sci.*, 7, 161, 1982.
20. Yamaguchi, I., Fungicides for control of rice blast disease, *J. Pest. Sci.*, 7, 307, 1982.
21. Suzuki, K., Kubo, Y., Furusawa, I., Ishida, N., and Yamamoto, M., Behavior of colorless appressoria in an albino mutant of *Colletotrichum lagenarium, Can. J. Microbiol.*, 28, 1210, 1982.
22. Homma, Y., Takahashi, K., Mizuno, H., and Misato, T., Studies on mode of action of lecithin. I. Effect on powdery mildew fungus of cucumber, *Ann. Phytopathol. Soc. Jpn.*, 40, 304, 1975 (Abstr. in Japanese).
23. Endo, T., Wakae, O., and Matsuura, K., Studies on the mode of action of Validamycin against leaf sheath-blight disease of rice. XIII. Anatomycal observations, *Ann. Phytopathol. Soc. Jpn.*, 41, 301, 1975 (Abstr. in Japanese).
24. Yamada, Y., Saito, J., and Takase, I., Development of a new fungicide, Pencycuron, *J. Pest. Sci.*, 13, 375, 1988.
25. Cardoso, J. E. and Echandri, E., Nature of protection of bean seedling from *Rhizoctonia* root rot by a binucleate *Rhizoctonia*-like fungus, *Phytopathology*, 77, 1548, 1987.
26. Gäumann, E., Neaf-Roth, S., and Miesher, G., Untersuchungen uber das Lycomarasmin, *Phytopathol. Z.*, 16, 257, 1950.
27. Kalyansundarem, R. and Saraswathi-Devi, L., Zinc in the metabolism of *Fusarium vasinfectum* Atk, *Nature (London)*, 175, 945, 1955.
28. Davis, D., Fusaric acid in selective pathogenicity of *Fusarium oxysporum, Phytopathology*, 59, 1391, 1969.
29. Sanwall, B. D., Investigation on the metabolism of *Fusarium lycopersici* Sacc. with the aid of radioactive carbon, *Phytopathol. Z.*, 25, 333, 1956.
30. Nishimura, S., Recent development of host-specific toxin research in Japan and its agricultural use, in *Molecular Determinant of Plant Diseases*, Nishimura, S., Vance, C., and Doke, N., Eds., Japan Scientific Society Press, Tokyo/Springer-Verlag, Berlin, 1987, 11.
31. Tusge, T., Nishimura, S., Omura, S., Kohmoto, K., and Otani, H., Metabolic regulation of host-specific toxin production in *Alternaria alternata* pathogens. II. Suppression of toxin production from germinating spores by chemical treatments, *Ann. Phytopathol. Soc. Jpn.*, 51, 277, 1985.

32. Tamari, K. and Kaji, J., Studies on the mechanism of injurious action of fusaric acid on plant growth 1–5, *J. Agric. Chem. Soc. Jpn.*, 26, 223, 295, 345, 349, 1952.
33. Tamari, K. and Kaji, J., Studies on the mechanism of injurious action of fusaric acid on plant growth 6–8, *J. Agric. Chem. Soc. Jpn.*, 27, 245, 249, 302, 1953.
34. Tamari, K. and Kaji, J., Studies on the mechanism of injurious action of fusaric acid on plants, *Jpn. J. Biochem.*, 41, 143, 1954.
35. Deuel, H., Über Storungen des Spurenelementhaushaltes der Pflanzen durch Welketoxine, *Phytopathol. Z.*, 21, 337, 1954.
36. Broun, R., Über Wirkungsweise und Umwandelungen der Fusarinsäure, *Phytopathol. Z.*, 39, 197, 1960.
37. Kern, H., Phytotoxins produced by *Fusaria*, in *Phytotoxins in Plant Diseases*, Wood, R. K. S., Ballio, A., and Graniti, A., Eds., Academic Press, London, 1972, 35.
38. Oku, H. and Nakanishi, T., A toxic metabolite from *Ascochyta fabae* having antibiotic activity, *Phytopathology*, 53, 1321, 1963.
39. Oku, H. and Nakanishi, T., Mode of action of an antibiotic ascochitine, with reference to selective toxicity, *Phytopathol. Z.*, 55, 1, 1966.
40. Kern, H. and Neaf-Roth, S., Zur Bildung phytotoxischer Farbstoffe durch Fusarien der Gruppe Martiella, *Phytopathol. Z.*, 53, 45, 1965.
41. Sinden, S. L. and Durbin, R. D., Glutamine synthetase inhibition: possible mode of action of wildfire toxin from *Pseudomonas tabaci*, *Nature (London)*, 219, 379, 1968.
42. Patil, S. S., Tam, L. Q., and Kolattukudy, P. E., Isolation and mode of action of the toxin from *Pseudomonas phaseolicola*, in *Phytotoxin in Plant Disease*, Wood, R. K. S. and Graniti, A., Eds., Academic Press, London, 1972, 365.
43. Hashimoto, H., Toyoda, H., and Ouchi, S., Role of fusaric acid in the symptom development in fusarial wilt of tomato, *Proc. UCLA Symp.*, 1988, Y232.
44. Anzai, H., Yoneyama, K., and Yamaguchi, I., Transgenic tobacco resistant to a bacterial disease by the detoxification of a pathogenic toxin, *Mol. Gen. Genet.*, 219, 492, 1989.
45. Akai, S., Studies on *Helminthosporium* blight of rice plant. VII. On the relation of silicic acid supply to the outbreak of *Helminthosporium* blight or blast disease in rice plants, *Ann. Phytopathol. Soc. Jpn.*, 17, 109, 1953.
46. Ishiyama, T. and Sato, K., Action of alkylthiocyanate compounds in control of rice blast disease, *Jubilee Publication in Commemoration of 60th Birthday of Professor M. Sakamoto*, 1968, 251.
47. Nasuda, K., Effects of spraying of some fungicides on the physiology of rice plants and on the increased rate of resistance against blast disease, *Spec. Bull. Fukui Agric. Exp. Stn.*, 3, 1, 1973.
48. Sempio, C., Contributo alla conoscenza del mecanismo della resistenza in dotta dal Cadmium nei tissuti del frumento, *Ann. Fac. Agrar. Regia Univ. Perugia*, 1, 3, 1942.
49. Meyer, H., Über den Einfluss von Cadmium auf die Krankheitsbereitschaft des Weizens fur *Erysiphe graminis tritici* Marchal, *Phytopathol. Z.*, 17, 63, 1951.
50. Müller, K. O., Mackay, H. E., and Fried, J. N., Effect of streptomycin on the host-pathogen relationship of a fungal pathogen, *Nature (London)*, 174, 878, 1954.

51. Vörös, J., Király, Z., and Farkas, G. L., Role of streptomycin-induced resistance to *Phytophthora* in potato, *Science*, 126, 1178, 1957.
52. Van Andel, O. M., Amino acids and plant diseases, *Annu. Rev. Phytopathol.*, 4, 349, 1966.
53. Akutsu, K., Amano, K., and Ogasawara, N., Inhibitory action of methionine upon the barley powdery mildew *(Erysiphe graminis* f. sp. *hordei)*. I. Microscopic observation of development of the fungus on barley leaves treated with methionine, *Ann. Phytopathol. Soc. Jpn.*, 43, 33, 1977.
54. Akutsu, K., Amano, K., and Yora, K., Inhibitory action of methionine upon the barley powdery mildew *(Erysiphe graminis* f. sp. *hordei)*. II. Electron microscopy of primary haustoria, hyphae and conidia on barley leaves treated with L-methionine, *Ann. Phytopathol. Soc. Jpn.*, 43, 537, 1977.
55. Watanabe, T., Igarashi, H., Matsumoto, K., Seki, S., Mase, S., and Sekizawa, Y., The characteristic of probenazole (Oryzemate) for control of rice blast, *J. Pest. Sci.*, 2, 291, 1977.
56. Sekizawa, Y. and Watanabe, T., On the mode of action of probenazole against rice blast, *J. Pest. Sci.*, 6, 247, 1981.
57. Langcake, P. and Wickins, S. G. A., Studies on the action of the dichlorocyclopropanes on the host-parasite relationship in rice blast disease, *Physiol. Plant Pathol.*, 7, 113, 1975.
58. Cartwright, D. W., Langcake, P., and Ride, J. P., Phytoalexin production in rice and its enhancement by a dichlorocyclopropane fungicide, *Physiol. Plant Pathol.*, 17, 257, 1980.
59. Ward, E. W. B., Lazarovits, G., Stoessel, P., Barrie, D., and Unwin, C. H., Glyceollin production associated with control of *Phytophthora* rot of soybeans by the systemic fungicide, metalaxyl, *Phytopathology*, 70, 738, 1980.
60. Király, Z., Barna, B., and Ersek, T., Hypersensitivity as a consequence, not the cause, of plant resistance to infection, *Nature (London)*, 234, 456, 1972.
61. Lazarovits, G. and Ward, E. W. B., Relationship between localized glyceollin accumulation and metalaxyl treatment in the control of *Phytophthora* rot in soybean hypocotyls, *Phytopathology*, 72, 1217, 1982.
62. Barak, E., Edington, L. V., and Ripley, B. D., Bioactivity of fungicide metalaxyl in potato tubers against some species of *Phytophthora, Fusarium*, and *Alternaria*, related to polyphenoloxidase activity, *Can. J. Plant Pathol.*, 6, 304, 1984.
63. Börner, H., Schatz, G., and Grisebach, H., Influence of the systemic fungicide metalaxyl on glyceollin accumulation in soybean infected with *Phytophthora megasperma* f. sp. *glycinea, Physiol. Plant Pathol.*, 23, 145, 1983.
64. Moesta, P. and Grisebach, H., L-2-Aminooxy-3-phenylpropionic acid inhibits phytoalexin accumulation in soybean with concomitant loss of resistance against *Phytophthora megasperma* f. sp. *glycinea, Physiol. Plant Pathol.*, 21, 65, 1982.
65. Ward, E. W. B., Suppression of metalaxayl activity by glyphosate: evidence that host defence mechanism contribute to metalaxyl inhibition of *Phytophthora megasperma* f. sp. *glycinea* in soybeans, *Physiol. Plant Pathol.*, 25, 381, 1984.
66. Keen, N. T., Holliday, M. J., and Yoshikawa, M., Effect of glyphosate on glyceollin production and the expression of resistance to *Phytophthora megasperma* f. sp. *glycinea, Phytopathology*, 72, 1467, 1982.

67. Cahill, D. M. and Ward, E. W. B., Effect of metalaxyl on elicitor activity, stimulation of glyceollin production and growth of sensitive and tolerant isolate of *Phytophthora megasperma* f. sp. *glycinea*, *Physiol. Mol. Plant Pathol.*, 35, 97, 1989.
68. Cohen, Y. and Coffey, M. D., Systemic fungicides and the control of Oomycetes, *Annu. Rev. Phytopathol.*, 24, 311, 1986.
69. Farih, A., Tsao, P. H., and Menge, J. A., Fungistatic activity of efosite aluminium on growth, sporulation, and germination of *Phytophthora parasitica* and *P. citrophthora*, *Phytopathology*, 71, 934, 1981.
70. Tey, C. C. and Wood, R, K. S., Effect of various fungicides in vitro on *Phytophthora palmivola* from cocoa, *Trans. Br. Mycol. Soc.*, 80, 271, 1983.
71. Fenn, M. E. and Coffey, M. D., Studies on the in vitro and in vivo antifungal activity of fosetyl-Al and phosphorus acid, *Phytopathology*, 74, 606, 1984.
72. Coffey, M. D. and Brower, L. A., In vitro viability among isolates of eight *Phytophthora* species in response to phosphorous acid, *Phytopathology*, 74, 738, 1984.
73. Fenn, M. E. and Coffey, M. K., Further evidence for the direct mode of action of fosetyl-Al and phosphorous acid, *Phytopathology*, 75, 1064, 1985.
74. Bower, L. A. and Coffey, M. D., Development of tolerance to phosphorous acid, fosetyl-Al, and metalaxyl in *Phytophthora capsici*, *Can. J. Plant Pathol.*, 7, 1, 1985.
75. Guest, D. I., Modification of defence responses in tobacco and *Capsicum* following treatment with fosetyl-Al, *Physiol. Plant Pathol.*, 25, 125, 1984.
76. Guest, D. I., Evidence from light microscopy of living tissues that fosetyl-Al modifies the defence response in tobacco seedlings following inoculation by *Phytophthora nicotianae* var. *nicotianae*, *Physiol. Plant Pathol.*, 29, 251, 1986.
77. Nemestothy, G. S. and Guest, D. I., Phytoalexin accumulation, phenylalanine ammonia lyase activity and ethylene biosynthesis in fosetyl-Al treated resistant and susceptible tobacco cultivars infected with *Phytophthora nicotianae* var. *nicotianae*, *Physiol. Mol. Plant Pathol.*, 37, 207, 1990.
78. Abu-Jawdah, Y. and Kummert, J., Effect of Aliette on AMV infection of bean leaves and on the resultant alterations in the patterns of proteins and peroxidases, *Phytopathol. Z.*, 108, 294, 1983.
79. Tomita K., Takahi, Y., Ishizuka, R., Kamimura, S., Nakagawa, M., Ando, M., Nakanishi, T., and Okudaira, H., Hymexazole, a new plant protecting agent, *Ann. Sankyo Res. Lab.*, 25, 1, 1973.
80. Nakanishi, T., Takahi, Y., and Tomita, K., Mode of action of hymexazole. I. Effect of hymexazole on cell permeability of *Pellicularia sasakii*, *Ann. Phytopathol. Soc. Jpn.*, 41, 321, 1975.
81. Kamimura, S., Nishikawa, M., and Takahi, Y., Mode of action of soil fungicide hymexazole, 3-hydroxy-5-methylisoxazole, on *Fusarium oxysporum* f. *cucumerinum*, *Ann. Phytopathol. Soc. Jpn.*, 42, 242, 1976.
82. Kamimura, S., Nishikawa, M., Saeki, H., and Takahi, Y., Absorption and metabolism of 3-hydroxy-5-methylisoxazole in plants and the biological activities of the metabolites, *Phytopathology*, 64, 1273, 1974.
83. Ogawa, M., Ota, Y., Yamazaki, Y., and Tomita, K., Plant growth regulating activities of 3-hydroxy-5-methylisoxazole. I. Effects of 3-hydroxy-5-methylisoxazole and metabolites in plants on growth of rice seedlings, *Proc. Crop. Sci. Soc. Jpn.*, 42(1) (Extra Issue), 61, 1973.

CHAPTER 5

Systemic Induced Resistance as a Tool for Disease Control

Vaccinations are very effective tools to protect vertebrates from infectious diseases. There are reports of trials 100 years ago to immunize plants against disease by the method of vaccination.[1] In ancient times, heat-killed pathogens, or culture filtrates of pathogenic microorganisms were used as inducers of resistance, just as antigens in vaccination of animals. Among these, there were examples of effective control of plant diseases, but the effects were not high enough for practical use. Further, because the control efficacy in plant diseases is not as specific as in the antigen-antibody reaction in animals, the interest of plant pathologists declined. Later, with the discovery of the interference phenomenon in plant virus diseases or phytoalexin biosynthesis in fungal plant diseases the attention of scientists was renewed; and at present, several effective control measures have been developed.

McKinney[2] found that tobacco infected by some strains of tobacco mosaic virus (TMV) became resistant to another strain of TMV, and he called this phenomenon interference. In this case, the induced resistance was local, namely, cells around the inoculated part of tobacco became resistant to another strain.

Later, many scientists found evidence that the part of the plant other than the infected portion became resistant to diseases not only in viral but also in bacterial and fungal diseases. These phenomena have been called acquired resistance. Nowadays, both local and systemic induction of resistance are called interference, cross protection, or acquired resistance by many scientists.

In many plants, the systemic resistance is induced by inoculating a part of the plant by compatible (virulent)[3–5] and also by incompatible (avirulent)[3,5–8] pathogens. Systemic resistance is sometimes induced only by wounding a part of the plant.[9–11]

The mechanism of systemic induced resistance has not yet been fully elucidated, but many plant pathologists have been involved in this problem. As a result, several very effective measures for plant disease control have been developed.

I. BIOLOGICAL CONTROL OF VIRUS DISEASES BY ATTENUATED VIRUS STRAINS

A. Citrus Tristeza Virus

Citrus tristeza virus (CTV) originated in Africa; it was introduced to South America in the 1920s and almost wiped out the citrus industry in Argentina, Brazil, Uruguay, and Java in the 1940s. The disease also became very important in California after the 1940s. One of the most successful practical uses of cross protection for control of this disease was developed in Brazil,[12] and at present, this method is widely used throughout the world.

In the beginning of the 1960s, an active search for a mild strain of CTV was made. That is, trees having good growth and general appearance were sought in citrus orchards where heavy disease occurred. It was considered that in such citrus trees, the pathogenicity of CTV had been attenuated. As a result, some 45 mild strains of CTV were obtained. Of these, three strains showed satisfactory cross protection against severe virus strains.

CTV-protected budwood on tristeza-tolerant root stocks were made available to citrus growers in 1968. To date, no evidence has been found that the protection has broken down.

B. TMV of Tomato

Since cross protection against the tobacco mosaic virus of greenhouse tomatoes by an attenuated TMV strain has become commercially successful in Japan, this method has been widely used in many countries.

Fundamental studies on mild or attenuated strains of TMV — their production and detailed nature — were done by Holmes.[13] He obtained a mild strain by using a masking phenomenon. That is, he heated the pieces of tomato stems inoculated with TMV and incubated them in loosely plugged, sterilized test tubes at 34.6°C for 15 days. The stems were crushed, and the expressed juices were used as the inoculum of mild strains. In this report, he described that the most pronounced effect in blocking the spread of the distorting-strain virus was found when the masked strain was introduced about 1 week before the introduction of the distorting strain.

Oshima et al.[14] obtained a mild TMV strain by the same method as Holmes and confirmed the strain has some protective effect. The strain was called

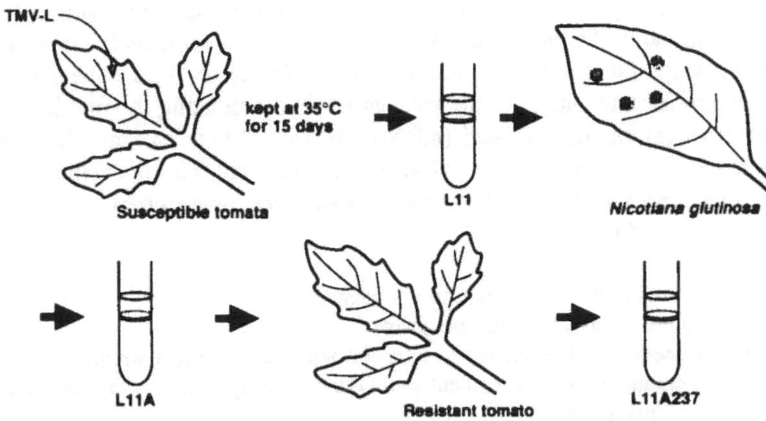

FIGURE 1. Preparation procedure of mild strain L_{11} and L_{11A237} of TMV for controlling TMV of tomato.

L_{11} as indicated in Figure 1. However, L_{11} showed slight mosaic symptoms on tomatoes, and the protective effect was not as prominent as in the practical use. Goto and Nemoto[15] inoculated L_{11} to *Nicotiana glutinosa*, a TMV-resistant tobacco, and isolated the TMV strain from the necrotic local lesion repeatedly. As a result, he finally obtained a stable, symptomless strain and named it L_{11A}. This strain has been distributed to growers to control TMV of the tomato. Later on, Oshima et al.[16] obtained strain L_{11A237} which can multiply in resistant tomato cultivars. Both strains, LA_{11A} and L_{11A237}, are now widely used in Japan.

In practical usage, tomato leaves carrying L_{11A} or L_{11A237} were homogenized and diluted with water, with 600–800 mesh carborundum added at a ratio of 20 g/L and sprayed on tomato seedlings of the 1–2 leaf stage from a distance of 5 cm at a pressure of 5–6 kg/cm^2. A good protective effect appears 18 days after inoculation with L_{11A}, and the effect continues at least for 1 month. The protection is more effective against the parent tomato L strain, than the tobacco OM strain.

According to large-scale experiments, the protected tomatoes both in greenhouses and in the field showed no symptoms until harvest. An increase in yield of 15–30% was obtained as compared with untreated tomatoes.

Nishiguchi and co-workers[17] determined and compared a full nucleotide sequence of L_{11A}-RNA with that of a virulent one, L; and found that 10 nucleotides were different from the parent L strain. Of these, changes of seven nucleotides do not affect the amino acid composition of the protein, but substitution of the other three nucleotides causes the change of amino acid composition. All these changes occurred between 130 and 180 kDa proteins. That is, 1117th, 2349th, and 2754th nucleotides from the 5′ terminal of viral RNA of the L strain, G, A, and G were substituted in the L_{11A} strain for A, G, and A, respectively. As a result, the corresponding amino acids, Cys, Asn, and Gly in L were substituted for Tyr, Asp, and Arg in L_{11A}, respectively. Of these, the mutation of 1117th G to A is believed to be necessary for attenuation of the virulent strain by reducing the multiplication of messenger ribonucleic acid (mRNA) for the 30-kDa protein which is responsible for the translocation of virus particles between host cells.

Several points we must be careful about when using cross protection to control virus disease are

1. Mutation of a mild strain to a virulent strain
2. Injurious effect on the host plant
3. Synergistic effect of mixed infection with other virulent strains
4. Contamination of a virulent strain during multiplication of a mild strain in host plants

To avoid these risks, the L_{11A} strain is multiplied in limited Agricultural Experiment Stations under strict supervision, and supplied to growers.

II. BIOLOGICAL CONTROL OF FUNGAL DISEASES BY AVIRULENT STRAINS

Although no commercial example is available for the control of fungal disease by cross protection at present, many trials have been made. Most of these experiments involve the use of *formae speciales* or races of pathogenic fungi as the inducer of resistance. This establishment of infection on plants is essential to induce resistance. That is, *formae speciales* or races, pathogens of other plant genera or cultivars, can establish infection on plants but cannot grow; and then they induce resistance. However, the use of *formae speciales* or races is not desirable because these fungi have pathogenicity on their respective host plants, that is, have some risks to nontarget plants.

Ogawa and Komada[18,19] developed a very effective measure to control wilt disease of the sweet potato caused by *Fusarium oxysporum* f. sp. *batatas* using a nonpathogenic isolate of the same fungus as the inducer of systemic resistance. These methods have been found to be effective against the *Fusarium* wilt disease of other crop plants, and now active research is being conducted for commercial use. The outline of these methods will be discussed.

Until recently, *Fusarium* wilt was not a serious disease of the sweet potato, but a new, very delicious cultivar Benikomachi was developed. Unfortunately, this cultivar was highly susceptible to *Fusarium* wilt, and an effective control measure was needed.

Ogawa and Komada[18,19] found that isolates of nonpathogenic *Fusarium oxysporum* were often isolated from tissues of healthy sweet potato sprouts and sometimes from healthy tubers; and these hyphae can also be observed microscopically in vascular bundles of the stem. These isolates were found

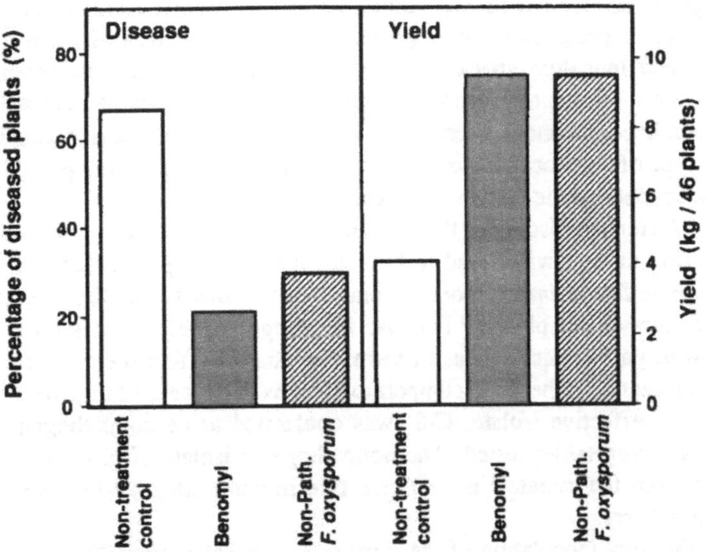

FIGURE 2. Effect of preinoculation with nonpathogenic *Fusarium oxysporum* on disease incidence of *Fusarium* wilt and yield of sweet potato in comparison with dipping the sprouts into benomyl suspension (500 times of 50% W.P.) in naturally infested field. (From Ogawa, K. and Komada, H., *Ann. Phytopathol. Soc. Jpn.*, 50, 5, 1984. With permission.)

to be nonpathogenic not only to sweet potatoes but also to all other plant species tested, such as the tomato, cucumber, melon, radish, and so on. Prior inoculation with some of these nonpathogenic isolates showed a good controlling effect against virulent isolates of *Fusarium oxysporum*. No antagonistic interaction was observed between nonpathogenic and pathogenic isolates in vitro. Therefore, they concluded that the control effect is due to the cross protection associated with the host response. Although the nonpathogenic isolate persisted locally at the cut end of the stem where it was inoculated and grew upward through the vessel, the entire stem as well as the cut end was protected against disease. Thus, the cross protection was proved to be systemic.

From these experimental results, they developed a very effective biocontrol measure for the *Fusarium* wilt of the sweet potato. In practice, the nonpathogenic isolate of *Fusarium oxysporum*, No. 5 or No. 101–2, was shake cultured for 5–7 days; mycelia were eliminated by filtration through cheesecloth; and the resultant filtrate, bud-cell suspension, was used as the inducer. The freshly cut surface of sprouts was dipped in the bud-cell suspension or treated with slurry of the concentrated bud-cell, and then these sprouts were planted on soil. As shown in Figure 2, the experimental results obtained from the naturally infested field were very good. This cross protection method gave nearly the same effect as chemical control with benomyl.

Ogawa and Komada found two kinds of responses related to the mechanism of this cross protection. First, germination and elongation of germ tubes of challenged inoculum spores are markedly inhibited in the protected sweet potato stem segment. Second, the protected segment became resistant to the phytotoxic compound(s) secreted by pathogenic isolates. Development of the same type of biocontrol measure is now in progress on the other plant diseases using nonpathogenic isolates of *Fusarium oxysporum*.

The *Fusarium* disease of the greenhouse strawberry is also a serious problem. This disease was found to be controlled by the preinoculation of nonpathogenic *Fusarium oxysporum* isolated from vascular bundles of crowns of healthy strawberry plants.[20] Because the nonpathogenic isolates do not compete with pathogenic isolates in vitro, the control effect seems to be due to cross protection rather than competition just as in the case of the sweet potato. The most effective isolate, C-8, was confirmed to be nonpathogenic to all crops and vegetables tested. The nonpathogenic isolate of *F. oxysporum* obtained from the tomato[21] is also effective in controlling *Verticillium* wilt of the strawberry.

Preliminary inoculation of the tomato with isolated nonpathogenic tomato *Fusarium oxysporum* was also found to reduce the disease incidence caused by *Verticillium dahliae*. In the protected tomato, peroxidase activity and concentration of phenolic compounds were increased by challenge inoculation with *V. dahliae*, just as the resistant cultivar was inoculated.

III. MECHANISMS OF SYSTEMIC INDUCED RESISTANCE

Loebenstein and van Praagh[22] found that inoculation of the lower part of a *Datura stramonium* leaf with TMV or tobacco necrosis virus (TNV) induced resistance in the upper uninoculated part of the same leaf to the same viruses. They isolated an "interferon-like substance" from the uninoculated part of the leaf. The substance inhibited virus infection and was found to be a protein with a molecular weight less than 30,000 Da. Research on the mechanism of systemic induced resistance has been conducted actively by Kuć and coworkers using the green bean, cucumber, tobacco, and so on; and has been reviewed by Kuć.[23–26]

Systemic resistance could be induced in the bean against anthracnose disease by inoculation with bean nonpathogen *Colletotrichum lagenarium* as well as by cultivar nonpathogenic race of bean pathogen *Colletotrichum lindemuthianum*. Although phytoalexin accumulation is associated with the race-specific resistance of bean cultivars to *C. lindemuthianum* as well as induced local protection by *C. lindemuthianum* or *C. lagenarium*, it is not sufficient to explain the systemic protection,[27] because phaseollin and other phytoalexins were not detected in systemically protected unchallenged tissue. Phytoalexin accumulation became evident only after the challenge inoculation. From these results, Kuć[24] concluded that at least two kinds of biologically active com-

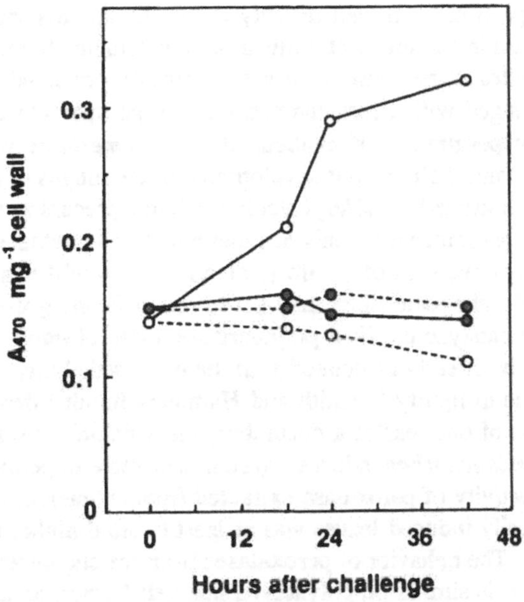

FIGURE 3. Lignin content of 'Marketer' cucumber apical tissue after challenge with *Cladosporium cucumerinum*. Lignin, extracted from cucumber cell walls with host aqueous alkali, was determined colorimetrically with diazotized-p-nitroaniline. -o- Systematic resistance induced with *Cladosporium lagenarium* and challenged with *C. cucumerinum;* -●- resistance not induced, tissue challenged; - -o- - resistance induced, tissue not challenged; - -●- - resistance not induced, tissue not challenged. (From Hammerschmidt, R. and Kuć, J., *Physiol. Plant Pathol.*, 20, 67, 1982. With permission.)

ponents are involved in induced systemic resistance: the chemical agent including phaseollin and other phytoalexins which accumulate around the infection site and contribute to the inhibition of fungal development; and the signal molecule that moves from the inoculation site and causes systemic resistance.

In the cucumber more than 12 pathogens — including viruses, bacteria, and fungi — have been shown to induce systemic resistance; and the resistance has been shown to be nonspecific with respect to the inducing pathogen and the challenge pathogen.[26]

As to the mechanism of resistance at the host-parasite interface in systemically resistance-induced plants, it is likely that several defense reactions are working in combination, since the resistance is displayed against fungi, bacteria, and viruses. Hammerschmidt and Kuć[28] and Dean and Kuć[29] reported that lignification of the host cell wall may be a mechanism for induced systemic resistance in the cucumber. That is, lignification occurred more rapidly and to a greater extent in protected cucumbers with *Colletotrichum lagenarium* than in the control after challenge with *Cladosporium cucumerinum* (Figure 3). The inhibition of penetration was associated with lignification of the host epidermal cell wall directly under the appressorium. The lignification reaction in the resistant cultivar was histologically indistinguishable from that at sites of penetration in a systemically protected "susceptible" cultivar challenged with *C. cucumerinum*. The germination of conidia and frequencies of penetration of challenged *C. cucumerinum* were similar in protected and control plants, but development of the fungus in lignified tissue was markedly restricted. Coniferyl alcohol, a lignin precursor, seemed to play some role as a resistance mechanism, functioning as a phytoalexin. They also found that the conversion of coniferyl alcohol to ligninlike material on mycelial walls of pathogenic fungi may act to retard fungal growth.

Peroxidases catalyze the final polymerization step of lignin synthesis, and may therefore be directly associated with the increased ability of systemically protected tissue to lignify.[30] Smith and Hammerschmidt[31] demonstrated that the inoculation of one leaf of a cucumber, muskmelon, or watermelon with *Colletotrichum lagenarium* induces a systemic increase in peroxidase activity. The specific activity of peroxidase extracted from an intercellular wash fluid from systemically induced leaves was at least twofold higher than that from control leaves. The behavior of peroxidases from the cucumber, muskmelon, and watermelon is similar on polyacrylamide gel; further, an antibody raised against purified cucumber peroxidase reacted with muskmelon and watermelon peroxidases of partial identity in the double diffusion assay.

Smith et al.[32] also found that inoculation of the first leaf of a cucumber with *Pseudomonas syringae* pv. *syringae* (wheat isolate) elicited a rapid hypersensitive response and a systemic resistance to *C. lagenarium*. In this report they described that the appearance of hypersensitive response elicited by bacterial plant pathogens is correlated with the rapidity of induction of systemic resistance and associated peroxidases. However, since removal of the inducer leaf prior to the appearance of visible necrosis still results in systemic induced resistance, signal(s) may not rely on cell death, but instead on early biochemical or physiological events which lead to hypersensitive reaction-related cell death. *Tn 5* mutants of *P. syringae* pv. *syringae*, selected for their inability to induce hypersensitive response and systemic peroxidase activity in the cucumber, had also lost their ability to elicit systemic resistance in the cucumber and cause disease on wheat. From these results they suggested that *P. syringae* pv. *syringae* may be useful in future studies on the nature of systemic signaling and the regulation of induced resistance in cucurbits.

There are reports[33,34] that one of the mechanisms for induced systemic resistance is the rapid and efficient formation of papillae. This phenomenon is due to the increased activity of the Ca^{2+}-regulated 1,3-β-glucan (callose) synthase.[34] This enzyme activity is latent in epidermal cells of unchallenged healthy cucumber leaves, but after activation renders the cell capable of a rapid production of callose-containing papillae. Activation occurs as soon as the Ca^{2+} permeability of the plasma membrane is perturbed by attempted fungal penetration.

Gessler and Kuć[35] obtained an active extract to induce systemic resistance from cucumber leaves which are infected by *C. lagenarium*, TMV, and *P. lachrymans*. Infiltration of active extracts into cotyledons and the first true leaf of the cucumber induced resistance to disease caused by *C. lagenarium*. A lag of 4–5 days was required between infiltration of the inducer and expression of the resistance. An extract of uninfected leaf tissue adjacent to the infection also elicits resistance when infiltrated to a susceptible plant; that is, the active principle may be transmitted into adjacent healthy tissues.

From the extract of infected leaves, they isolated a protein which has a molecular weight of ca. 16,000 Da. The protein is stable in acidic buffers, and is not detected in the extract of leaves which are damaged by physical treatment. The appearance of the protein coincides with that of the symptoms, e.g., 3–4 days after inoculation with *C. lagenarium*. At that time, it was not clear whether this protein is a pathogenesis-related (PR) protein, but it seemed to be one even though it markedly increased in the protected tissue after challenge. Later, Kuć's group (Ye et al.[36]) found that stem injection with sporangiospores of *Peronospora tabacina* or inoculation of leaves with TMV induced systemic protection in tobacco against the same fungus and virus. They also identified PR proteins from the extract of protected and uninfected plants and found that the appearance of PR proteins coincided with the appearance of protection. PR proteins accumulated earlier and to a higher level in the protected plant after challenge with the TMV or *P. tabacina* than in the control. Some of the PR proteins were detected in intercellular fluids of the induced plant before challenge. The pattern of PR proteins appeared qualitatively identical following induction with *P. tabacina* and TMV.

Ye et al.[37] also reported that PR proteins and activities of β-1,3-glucanase are associated with systemic induced resistance to *P. tabacina* but not to TMV. That is, a tobacco cultivar carrying the *N* gene is resistant to TMV producing local lesions (hypersensitive reaction) at 23°C but is susceptible (systemic mosaic) at 28°C or above. PR proteins and activities of peroxidase, β-1,3-glucanase, and chitinase increased systemically in the inoculated plants at 23°C. When TMV-inoculated leaves were removed 12 days after inoculation at 23°C and plants were challenged with TMV or *P. tabacina* and held at 23°C, induced resistance was apparent against both TMV and *P. tabacina*. At 28°C, induced tobacco responded to challenge inoculation with TMV or *P. tabacina* by producing more PR proteins; activities of peroxidase, β-1,3-glucanase, and chitinase; and resistance to *P. tabacina* but not to TMV. These results indicate that PR proteins and enzyme activities are responsible for induced resistance to *P. tabacina* but not to TMV.

Pan et al.[38] further reported that five isoforms of β-1,3-glucanase increased in the systemically protected tobacco plants by stem injection with *P. tabacina* or leaf inoculation with TMV; and these isoforms were designated as G1, G2, G3, G4, and G5. The activities of G1 and G2 were higher in leaves of systemically protected plants. G3 was not associated with protection. They compared these isoforms on polyacrylamide gel electrophoresis (PAGE) gels with 10 PR proteins previously reported in tobacco. As a result, G1 co-migrated with PR-N (2b); G2 co-migrated with PR-O (2c); and G3 did not co-migrate with any of the 10 PR proteins or four β-1,3-glucanases reported previously. G5 also did not co-migrate with any of the 10 PR proteins or four β-1,3-glucanases previously reported in tobacco, but this isoform was systemically induced in the TMV-induced plants. G4 appeared at the time of symptom expression in the control plant and seemed to be produced by pathogenic fungus *P. tabacina* and to be associated with symptom expression. The role of these isoforms in the induced resistance is not fully understood and hopefully will be elucidated.

Métraux and Bollar[39] found that the activity of chitinase, a fungitoxic enzyme, was induced in cucumber plants both locally and systemically in response to a localized infection by TNV, *Colletotrichum lagenarium*, *Pseudomonas lachrymans*, and *Pseudoperonospora cubensis*. In the infected area of the leaf, the increase in activity ranged from 60- to 2000-fold. In case of fungal infection, some of the increases are thought to come from fungal chitinase, but a similar increase was also observed by infection with TNV or *Pseudomonas lachrymans;* hence increased chitinase might be of host origin.

Chitinase activity also increased up to 100-fold in uninfected second leaves of the cucumber plant of which the first leaf was infected. There was a positive correlation between the level of systemic induced resistance and the activity of chitinase. That is, the induction with *Pseudoperonospora cubensis* led to the highest activity in chitinase and to the best protection against *C. lagenarium*. They also induced the systemic increase in chitinase and resistance by treatment with abiotic agents, ethylene and necrotizing salt solutions. In these cases, the chitinase activity increased 20- to 50-fold in the first leaves and 3- to 6-fold in the untreated second leaves. Thus they suggested the importance of the induction of chitinase as a mechanism of induced resistance.

As described in Chapter 2, four PR proteins were recently identified as chitinases. In fact, Métraux et al.[40] identified a PR protein induced in cucumber leaves by inoculation with fungal, bacterial, or viral pathogens as chitinase. That is, they found a PR protein in both the challenged and unchallenged leaves of cucumber plants inoculated on the first true leaf with a fungal, a bacterial, or a viral pathogen. This host-coded protein was detected up to five leaves above the infected leaf. The protein was purified and identified as chitinase with a molecular weight of 28,000 Da. As to the mode of action of chitinase in resistance, they consider two possibilities; one is that chitinase inhibits fungal growth directly hydrolyzing the fungal wall, and the other possibility is that chitinase releases endogenous elicitors which then trigger defense reactions in the host and reduce or prevent invasion by the pathogen. Bollar and Métraux[41] demonstrated that most of the chitinase activity is located in the extracellular space (intercellular fluid).

IV. BIOTIC SIGNAL MOLECULES FOR INDUCTION OF SYSTEMIC RESISTANCE

Resistance is considered to be induced systemically by some endogenous signal molecules produced in the infection sites, which are translocated to other parts of the plants. The search and identification of the putative signal for induced resistance are the current interest of many scientists, since such molecules have possible uses as disease control agents.

Métraux et al.[42] analyzed phloem exudates of the cucumber and found that the content of salicylic acid correlates with the degree of systemic induced resistance. Changes of salicylic acid content were monitored after infection of the first leaf either by *C. lagenarium* or TNV, and it was found that an increase in salicylic acid up to 10-fold was detected in the phloem 1 day before resistance was induced in the upper leaf. Because salicylic acid is not metabolized into fungitoxic compounds in the cucumber, salicylic acid is considered to be the putative signal. Malamy et al.[43] obtained similar results and found that the level of endogenous salicylic acid increased in protected tobacco leaves in parallel to the appearance of PR proteins.

Bollar et al.[44] reported that the response of a suspension cultured tomato cell to an elicitor (prepared from yeast extract) is mediated by ethylene which is produced after an elicitor treatment.

Inoculation of barley leaves with powdery mildew fungus, *Erysiphe graminis* f. sp. *hordei*, is known to induce systemic resistance against a compatible race of the same fungus, regardless of the compatibility of the inducer race.[5,45] Recently, systemic resistance has been found to be induced very rapidly.[5] That is, the secondary leaves of seedlings of which primary leaves were inoculated with a compatible or an incompatible race of powdery mildew fungus were detatched from the seedlings at time intervals after inoculation, supplied with water from the cut end of leaves, and then challenged with a compatible race to determine the degree and the exact time required for induced systemic resistance. As a result, it was found that the time required for induction of systemic resistance is only 1.5 hr after the inoculation of the primary leaf (Figure 4). The mechanism of systemic resistance induced in the barley plant seems to be different from the cucumber-anthracnose system since 3 days are needed for resistance induction. Freeze-killed conidia of the powdery mildew fungus do not induce resistance in barley seedlings. The resistance induced in barley leaves is only effective in inhibiting the infection frequency but not in inhibiting the elongation of secondary hyphae.

The mechanism of systemic resistance in barley seedlings induced by powdery mildew fungi involves two possibilities; one is that the signal(s) to induce resistance originates from inducer fungus living conidia; and the other is that the mechanical stimuli elicits the signal(s). To elucidate these two possibilities, Fujiwara et al.[10] studied the effect of injury, pruning of organs, on the induction

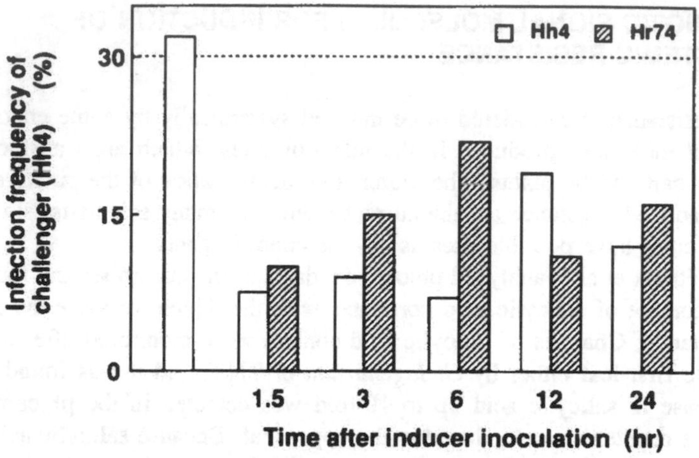

FIGURE 4. Rapid induction of systemic resistance in secondary leaves of barley cultivar H.E.S. 4 by preliminary inoculation of primary leaves with compatible (Hh4) and incompatible (Hr74) races of powdery mildew fungus. (Modified from Fujiwara, M. et al., *Ann. Phytopathol. Soc. Jpn.*, 55, 660, 1989.)

of systemic resistance. As a result, pruning of the primary leaf induced in the second leaf of the barley seedling resistance to infection by the powdery mildew fungus. Pruning of the secondary leaf also induced resistance in the primary leaf. That is, no direction was found in the transmission of the signal(s) responsible for induced systemic resistance. The pruning of the root and the remaining albumen also induced resistance. The resistance was induced very rapidly after pruning, within 2 hr, and continued for at least 6 hr. It is interesting that pruning of the root with a hypocotyl induced no more resistance. This fact suggests the importance of the hypocotyl in induction of systemic resistance. The hypocotyl may play a role in the transfer of the signal(s) to the remaining leaves. In the primary leaf, a susceptibility was induced apparently in all experiments of periodic pruning of roots with hypocotyls. This result also suggests the possible importance of the hypocotyl in maintaining resistance in this host-parasite combination.

Fujiwara et al.[11] revealed further that the incubation of barley seedlings in a container with seedlings on which primary leaves were pruned induced in the intact seedlings a systemic resistance to the powdery mildew fungus. The induced resistance was more prominent in the primary leaf than in the secondary leaf. These results suggest that some volatile substance(s) which is released after pruning of leaves may be responsible for the induced systemic resistance. The decline of resistance with time indicates that the volatile substance may be released at once by the pruning of leaves and then decreases. The primary leaves seem to be more sensitive than the secondary leaves to the volatile substance. The volatile substance trapped by cooling with liquid nitrogen was examined but has not yet been identified; however, neither

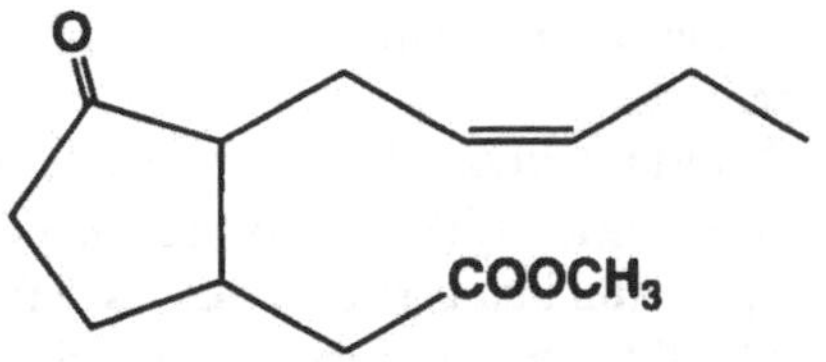

FIGURE 5. Methyl jasmonate.

ethylene nor salicylic acid were identified by Métraux et al.[42] or Bollar et al.[44] from a suspension of cultured tomato or cucumber cells.

A systemic resistance induced in the secondary leaves of the same seedlings on which primary leaves were pruned continued for 120 hr; therefore many substances (at least other than volatile substances) may be involved in the induction of systemic resistance.

Arachidonic acid and eicosapentaenoic acid are known to be elicitors for sesquiterpenoid phytoalexins in the potato produced by the late blight fungus, *Phytophthora infestans*.[46] Cohen et al.[47] reported that arachidonic acid, eicosapentaenoic acid, and several other unsaturated fatty acids have the ability to induce systemic resistance in potato plants to the late blight fungus. That is, arachidonic and eicosapentaenoic acid applied to leaves 1–3 of potato plants at a dose of about 1 mg per plant induced 94–97% protection in leaves 4–11, respectively. Linoleic acid, linolenic acid, and oleic acid provided 82, 39, and 42% protection, respectively. Protection was evident as a reduction of the lesion number and size. The efficacy of protection was correlated with the phytotoxicity of these unsaturated fatty acids. As the mechanism of protection, they considered that the actual signal compound(s) is not these fatty acids but a metabolite(s) possibly produced by the plant as a direct effect of fatty acids on the plasma membrane. Although these fatty acids are known to be elicitors of sesquiterpenoid phytoalexins in the potato as described, they could not succeed in extracting these stress compounds from treated plants. The systemic protection in the potato by treatment with a hyphal cell wall preparation of the late blight fungus[48] remains a possibility in that the protection may be caused by arachidonic acid and eicosapentaenoic acid contained in the fungal wall.[46]

Methyl jasmonate (Figure 5), a common secondary metabolite of plants, has been reported to be a signal molecule of systemic and local induced resistance produced at the site of attack by pathogens or insects.[49] When methyl jasmonate was applied to the surface of tomato plants, it induced the synthesis of defensive proteinase inhibitor proteins in treated plants and also in nearby plants. The incubation of tomato, tobacco, and alfalfa plants in the chamber containing methyl jasmonate resulted in the accumulation of proteinase inhibitors. From these results, methyl jasmonate is considered to be one of the interplant communication molecules. The airborne methyl jasmonate molecule is considered to enter the vascular system by way of stomata and to activate the proteinase inhibitor genes through a receptor-mediated signal transduction pathway.

Farmer and Ryan[50] further reported that octadecanoid precursors of jasmonic acid, i.e., linolenic acid, 13(S)-hydroperoxylinodenic acid, and phytodienoic acid, act as powerful inducers of proteinase inhibitors. By contrast, compounds closely related to the precursors but which are not intermediates in the jasmonic acid biosynthetic pathway did not induce the synthesis of proteinase inhibitors.

Systemic resistance to diseases is also induced by physical stresses through the production of abnormal metabolites which act a signal molecules. Keyworth and Dimond[9] reported that certain types of injury such as dipping in hot water, periodic pruning, and application of the sublethal doses of damaging compounds to the root of tomato plants before inoculation with *Fusarium oxysporum* f. sp. *lycopersici* result in a great reduction of disease severity. Removing the injured root immediately before inoculation did not affect the protective effect. From these facts, they emphasized that for evaluating the effect of agrochemicals on the tomato *Fusarium* wilt, the toxic effect on the root must be taken into account in considering the mode of action.

Symptoms of *Fusarium* wilt in tomatoes were reduced by treating the roots with hot water.[51] The most effective heat treatment was 12 and 24 hr before the inoculation at the temperature which gives visible and irreversible cell injury. The heat treatment of the aerial part also induces systemic protection, but it is not necessary to cause visible permanent damage. Therefore, the mechanism of systemic induced resistance in root and aerial parts might be different.

Cucumber seedlings, heat shocked for 40 sec at 50°C acquire resistance to the pathogen, *Cladosporium cucumerinum*.[52] The resistance develops 15–21 hr after heat shock treatment and is associated with ethylene production and changes in cell wall components. There was a twofold increase in ethylene by 6 hr after heat shock. Heat shock also enhanced the accumulation of insoluble extensin, a hydroxyproline-rich glycoprotein in cell walls of cucumber seedlings. The hydroxyproline-rich glycoprotein content of the cell wall of heat shocked seedlings increases after inoculation with *C. cucumerinum*. Peroxidase activity increased by heat shock or treatment with ethylene. Because extensin, a rod-shaped hydroxyproline-rich glycoprotein is slowly insolubilized in the cell wall by the peroxidase-mediated formation of isodityrosine cross-links, an increase in the cross-linking glycoprotein by peroxidase activity may be responsible for the induced resistance of cell walls from digestion by enzymes produced by *C. cucumerinum* and hence provides resistance to fungal penetration.[53]

Thus, the signal substance involved in the systemic induced resistance differs from plant to plant. Therefore, the search for these signal molecules having strong activity in various host-parasite combinations and devices for use as the disease control agent might be one of the strategies for plant disease control.

V. ABIOTIC AND SYNTHETIC SIGNAL COMPOUNDS THAT INDUCE SYSTEMIC RESISTANCE

The use of synthetic signal compounds as the inducers of systemic resistance might have been described in Chapter 4 (which treated the increase in resistance of cultivated plants) because it might be that, in some chemicals, the mechanism of resistance induction is similar. However, in this chapter the abiotic chemical compounds will be discussed from the basic standpoint of the induced systemic resistance.

As described before, Métraux et al.[42] and Malamy et al.[43] proposed the hypothesis that salicylic acid may play a role as a signal molecule in induced resistance in cucumbers. Spraying extracts of leaves of spinach or rhubarb was reported to induce systemic resistance in the cucumber against anthracnose disease caused by *Colletotrichum lagenarium*,[54] and the active component in the leaf extract was identified as oxalate. Dimethyl and diethyl esters of oxalic acid, malonate, formate, glycolate, acetate, fumarate, phthalate, and maleate were inactive; whereas ascorbate and glyoxalate were less active than oxalate. The concentration of oxalate sprayed on leaves did not inhibit the spore germination or mycelial growth of anthracnose fungus in vitro; hence, the resistance appeared to be a host-mediated response. The induced systemic resistance by oxalate was evident 20–36 hr after spraying.

Irving and Kuć[55] reported that spraying of 50 m*M* K_2HPO_4 to the first true leaf of the cucumber induced systemic resistance to anthracnose disease 2 to 7 days later. Within 16 hr after application, chitinase and peroxidase activities increased in induced leaves. Increased peroxidase activity alone does not correlate with the resistance to anthracnose, but they consider that if its activity is coupled with increase in chitinase, then induced resistance occurs.

Ye et al.[36] reported that spraying of a 0.05% solution of acetylsalicylic acid (ASA) on tobacco leaves induced resistance only in the treated leaves (local resistance) against TMV and blue mold with the concomitant induction of PR proteins. However, the repeated stem injection of ASA induced significant protection against TMV, but not against blue mold. In this case, only trace amounts of one of the PR proteins was induced in protected leaves. That is, no correlation was found between protection against TMV and PR protein in ASA-injected tobacco. Thus, PR proteins may not be responsible for induced protection against TMV.

Because several PR proteins were identified as β-1,3-glucanase,[56] PR proteins may play a role in resistance to *Peronospora tabacina* by degrading cell walls. In contrast, Ohashi et al.[57] treated leaves of tobacco cultivars, Samsun, Samsun NN, Xanthi, and Xanthi NN with salicylic acid under various conditions and concluded that the level of TMV multiplication or spread was closely linked with the amount of PR proteins in leaves. However, they produced transgenic tobacco plants with PR 1a/GUS genes which constantly produce PR 1a protein; but they found these transgenic plants did not show resistance to TMV,[58] though the synergistic or additive effect between PR proteins is still obscure. Treatment of tobacco suspension cultured cells with chemicals such as potassium salicylate, eosin, gibberellin A_3, and indoleacetic acid was also found to induce PR proteins.[59] PR proteins accumulated both in cells and culture medium.

Métraux et al.[60] found that 2,6-dichloro-isonicotinic acid and its ester derivative induce local and systemic resistance in the cucumber against *C. lagenarium* and other pathogens. These compounds have no direct fungicidal activity. The extract of plants treated with these compounds does not show any antifungal activity; that is, these compounds are not metabolized in plants to a fungitoxic compound. They compared the time course of resistance development by treatment with these compounds with that by inoculation with TNV on lower leaves. Both treatments induce resistance and chitinase activity on upper untreated leaves. The induction of systemic resistance and chitinase is faster and stronger by treatment with 2,6-dichloro-isonicotinic acid than TNV inoculation. Chitinase activity induced both by 2,6-dichloro-isonicotinic acid and by TNV results from the accumulation of chitinase-mRNA in the tissue. These compounds are not only effective against the anthracnose disease of cucumbers but also against several bacterial and fungal diseases of other plants. Both foliar and drench treatments are effective.

Thus, the search for and use of substances acting as the signal for induced systemic resistance may be one of the most effective strategies for developing disease control agents without pollution.

REFERENCES

1. Chester, K. S., The problem of acquired physiological immunity in plants, *Q. Rev. Biol.*, 8, 192, 1933.
2. McKinney, H. H., Mosaic diseases in the Canary Island, West Africa and Gibraltar, *J. Agric. Res.*, 39, 557, 1929.
3. Cruickshank, I. A. M. and Mandryk, M., The effect of stem infestation of tobacco with *Peronospora tabacina* Adam on foliage reaction to blue mold, *J. Aust. Inst. Agric. Sci.*, 26, 369, 1960.
4. Kuć, J., Shockley, G., and Kerney, K., Protection of cucumber against *Colletotrichum lagenarium* by *Colletotrichum lagenarium*, *Physiol. Plant Pathol.*, 7, 195, 1975.
5. Fujiwara, M., Shiraishi, T., Oku, H., Yamada, T., and Ouchi, S., Rapid induction of systemic resistance in barley against powdery mildew fungus by the preliminary inoculation with *Erysiphe graminis*, *Ann. Phytopathol. Soc. Jpn.*, 55, 660, 1989.
6. Elliston, J., Kuć, J., and Williams, E., Induced resistance to anthracnose at a distance from the site of the inducing interaction, *Phytopathology*, 61, 1110, 1971.
7. Jenns, A. and Kuć, J., Localized infection with tobacco necrosis virus protects cucumber against *Colletotrichum lagenarium*, *Physiol. Plant Pathol.*, 11, 207, 1977.

8. Ross, A. F., Systemic resistance induced by localized virus infections in beans and cowpeas, *Phytopathology*, 54 (Abstr.), 1436, 1964.
9. Keyworth, W. G. and Dimond, A. E., Root injury as a factor in the assessment of chemotherapeutant, *Phytopathology*, 42, 311, 1952.
10. Fujiwara, M., Oku, H., Shiraishi, T., and Ouchi, S., Induction of systemic resistance in barley leaves by pruning of organs against powdery mildew fungus, *Ann. Phytopathol. Soc. Jpn.*, 52, 330, 1986.
11. Fujiwara, M., Oku, H., and Shiraishi, T., Involvement of volatile substances in systemic resistance of barley against *Erysiphe graminis* f. sp. *hordei* induced by pruning of leaves, *J. Phytopathol.*, 120, 81, 1987.
12. Costa, A. S. and Müller, G. W., Tristeza control by cross protection a U.S.-Brazil cooperative success, *Plant Dis.*, 64, 538, 1980.
13. Holmes, F. O., A masked strain of tobacco-mosaic virus, *Phytopathology*, 24, 845, 1934.
14. Oshima, N., Komochi, S., and Goto, T., Virus disease control with mild vaccine. I. Control of tomato mosaic disease, *Annu. Rep. Hokkaido Agric. Exp. Stn.*, 83, 87, 1964.
15. Goto, T. and Nemoto, M., Virus disease control by attenuated virus strains — Selection of stable mild virus strain and the effect to other species of plants, *Annu. Rep. Hokkaido Agric. Exp. Stn.*, 99, 67, 1971.
16. Oshima, N., Osawa, T., Morita, H., and Mori, K., A new attenuated virus L_{11} A237, *Ann. Phytopathol. Soc. Jpn.*, 44, 504, 1978.
17. Nishiguchi, M., Kikuchi, S., Kiho, Y., Ohno, T., Meshi, T., and Okada, Y., Molecular basis of plant virulence; the complete nucleotide sequence of an attenuated strain of tobacco mosaic virus, *Nucl. Acids Res.*, 13, 5585, 1985.
18. Ogawa, K. and Komada, H., Biological control of *Fusarium* wilt of sweet potato by non-pathogenic *Fusarium oxysporum*, *Ann. Phytopathol. Soc. Jpn.*, 50, 1, 1984.
19. Ogawa, K. and Komada, H., Induced resistance to *Fusarium* wilt by nonpathogenic *Fusarium oxysporum* in sweet potato, *Abstr. Pap. 2nd MAFF Homeostasis Workshop, Microbial Antagonism, Targeting Biol. Control Plant Dis.*, National Institute Agro-Environmental Science, Ministry of Agriculture, Forestry and Fisheries of Japan, 1990, 16.
20. Tezuka, N. and Makino, T., Biological control of *Fusarium* wilt of strawberry by nonpathogenic *Fusarium oxysporum*, *Abstr. Pap. 2nd MAFF Homeostasis Workshop, Microbial Antagonism, Targeting Biol. Control Plant Dis.*, National Institute Agro-Environmental Science, Ministry of Agriculture, Forestry and Fisheries of Japan, 1990, 13.
21. Amemiya, Y., Induced resistance to *Verticillium* wilt in tomato by nonpathogenic *Fusarium oxysporum*, *Abstr. Pap. 2nd MAFF Homeostasis Workshop, Microbial Antagonism, Targeting Biol. Control Plant Dis.*, National Institute Agro-Environmental Science, Ministry of Agriculture Forestry and Fisheries of Japan, 1990, 15.
22. Loebenstein, G. and van Praagh, T., Extraction of a virus interfering agent induced by localized and systemic infection, in *Host-Parasite Relations in Plant Pathology*, Király, Z. and Ubrizsy, G., Eds., Research Institute of Plant Protection, Budapest, 1964, 53.

23. Kuć, J., Multiple mechanisms, reaction rate, and induced resistance, in *Plant Disease Control, Resistance and Susceptibility*, Staples, R. C. and Toennies-sen, G. H., Eds., Wiley-Interscience, New York, 1981, 259.
24. Kuć, J., Induced immunity to plant disease, *BioScience*, 32, 854, 1982.
25. Kuć, J., The immunization of cucurbits against fungal, bacterial, and viral diseases, in *Plant Infection, The Physiological and Biochemical Basis*, Asada, Y., Bushnell, W. R., Ouchi, S., and Vance, C. P., Eds., Japan Scientific Press, Tokyo/Springer-Verlag, Berlin, 1982, 137.
26. Kuć, J., Induced systemic resistance in plants to diseases caused by fungi and bacteria, in *The Dynamics of Host Defense*, Bailey, J. and Deverall, B., Eds., Academic Press, Sydney, 1983, 191.
27. Ellisstone, J., Kuć, J., Williams, E., and Rahe, J., Relation of phytoalexin accumulation to local and systemic protection of bean against anthracnose, *Phytopathol. Z.*, 88, 114, 1977.
28. Hammerschmidt, R. and Kuć, J., Lignification as a mechanism for induced systemic resistance in cucumber, *Physiol. Plant Pathol.*, 20, 61, 1982.
29. Dean, R. A. and Kuć, J., Rapid lignification in response to wounding and infection as a mechanism for induced systemic protection in cucumber, *Physiol. Mol. Plant Pathol.*, 31, 69, 1987.
30. Gross, G. G., Recent advances in chemistry and biochemistry of lignin, *Recent Adv. Phytochem.*, 12, 77, 1979.
31. Smith, J. and Hammerschmidt, R., Comparative study of acidic peroxidases associated with induced resistance in cucumber, muskmelon and watermelon, *Physiol. Plant Pathol.*, 33, 255, 1988.
32. Smith, J. A., Hammerschmidt, R., and Fulbright, D. W., Rapid induction of systemic resistance in cucumber by *Pseudomonas syringae* pv. *syringae*, *Physiol. Mol. Plant Pathol.*, 38, 223, 1991.
33. Stum, D. and Gessler, C., Role of papillae in cucumber leaves in the induced systemic resistance of cucumbers against *Colletotrichum lagenarium*, *Physiol. Mol. Plant Pathol.*, 29, 405, 1986.
34. Schmele, I. and Kauss, H., Enhanced activity of the plasma membrane localized callose synthese in cucumber leaves with induced resistance, *Physiol. Mol. Plant Pathol.*, 37, 221, 1990.
35. Gessler, C. and Kuć, J., Appearance of a host protein in cucumber plants infected with virus, bacteria and fungi, *J. Exp. Bot.*, 33, 58, 1982.
36. Ye, X. S., Pan, S. Q., and Kuć, J., Pathogenesis-related proteins and systemic resistance to blue mould and tobacco mosaic virus induced by tobacco mosaic virus, *Peronospora tabacina* and aspirin, *Physiol. Mol. Plant Pathol.*, 35, 161, 1989.
37. Ye, X. S., Pan, S. Q., and Kuć, J., Association of pathogenesis-related proteins and activities of peroxidase, β-1,3-glucanase and chitinase with systemic induced resistance to blue mould of tobacco but not to systemic tobacco mosaic virus, *Phys. Mol. Plant Pathol.*, 36, 523, 1990.
38. Pan, S. Q., Ye, X. S., and Kuć, J., Association of β-1,3-glucanase activity and isoform pattern with systemic resistance to blue mould in tobacco induced by stem injection with *Peronospora tabacina* or leaf inoculation with tobacco mosaic virus, *Phys. Mol. Plant Pathol.*, 39, 25, 1991.

39. Métraux, J. P. and Bollar, Th., The local and systemic induction of chitinase in cucumber plants in response to viral, bacterial and fungal infections, *Physiol. Mol. Plant Pathol.*, 28, 161, 1986.
40. Métraux, J. P., Streit, L., and Staub, Th., A pathogenesis-related protein in cucumber is a chitinase, *Physiol. Mol. Plant Pathol.*, 33, 1, 1988.
41. Bollar, T. and Métraux, J. P., Extracellular localization of chitinase in cucumber, *Physiol. Mol. Plant Pathol.*, 33, 11, 1988.
42. Métraux, J. P., Singer, H., Ryals, J., Ward, E., Wyss-Benz, M., Gaudin, J., Rashdorf, K., Schmid, E., Blum, W., and Inverardı, B., Increase in salicylic acid at the onset of systemic acquired resistance in cucumber, *Science*, 250, 1004, 1990.
43. Malamy, J., Carr, J. P., Klessig, D., and Raskin, I., Salicylic acid: a likely endogenous signal in the resistance response of tobacco to virus infection, *Science*, 250, 1002, 1990.
44. Bollar, T., Felix, J., Basse, C., Grosskopf, D., Regenass, M., Reinhaldt, D., and Spann, P., Signal transduction in response of plants to pathogens, *Abstr. Symp. Phytoalexin Beyond*, Dannenfels, Germany, 1991, Session 4–5.
45. Hwang, B. K. and Heitefuss, R., Induced resistance of spring barley to *Erysiphe graminis* f. sp. *hordei*, *Phytopathol. Z.*, 103, 41, 1982.
46. Bostock, R. M., Kuć, J., and Laine, R. A., Eicosapentaenoic and arachidonic acids from *Phytophthora infestans* elicit fungitoxic sesquiterpenoids in potato, *Science*, 212, 67, 1981.
47. Cohen, Y., Gisi, U., and Mosinger, E., Systemic resistance of potato plants against *Phytophthora infestans* induced by unsaturated fatty acids, *Physiol. Mol. Plant Pathol.*, 38, 355, 1991.
48. Doke, N., Ramirez, A. V., and Tomiyama, K., Systemic induction of resistance in potato plants against *Phytophthora infestans* by local treatment with cell wall components of the fungus, *J. Phytopathology*, 119, 232, 1987.
49. Farmer, E. E. and Ryan, C. A., Interplant communication: airborne methyl jasmonate induces synthesis of proteinase inhibitors in plant leaves, *Proc. Natl. Acad. Sci. U.S.A.*, 87, 7713, 1990.
50. Farmer, E. E. and Ryan, C. A., Octadecanoid precursors of jasmonic acid activate the synthesis of wound-inducible proteinase inhibitors, *Plant Cell*, 4, 129, 1992.
51. Anchisi, M., Gennari, M., and Matta, A., Retardation of *Fusarium* wilt symptoms in tomato by pre- and post-infection treatments of the roots and aerial parts of the host in hot water, *Physiol. Plant Pathol.*, 26, 175, 1985.
52. Stermer, B. A. and Hammerschmidt, R., Association of heat shock induced resistance to disease with increased accumulation of insoluble extensin and ethylene synthesis, *Physiol. Mol. Plant Pathol.*, 31, 453, 1987.
53. Cooper, J. B. and Varner, J. E., Cross-linking of soluble extensin in isolated cell wall, *Plant Physiol.*, 76, 414, 1984.
54. Doubrava, N. S., Dean, R. A., and Kuć, J., Induction of systemic resistance to anthracnose caused by *Colletotrichum lagenarium* in cucumber by oxalate and extracts from spinach and rhubarb leaves, *Physiol. Mol. Plant Pathol.*, 33, 69, 1988.
55. Irving, H. R. and Kuć, J., Local and systemic induction of peroxidase, chitinase and resistance in cucumber plants by K_2HPO_4, *Physiol. Mol. Plant Pathol.*, 37, 355, 1990.

56. Kaufman, S., Legrand, M., Geoffroy, P., and Fritig, B., Biological function of pathogenesis-related proteins: four PR-proteins of tobacco have 1,3-glucanase activity, *EMBO J.*, 6, 3209, 1987.
57. Ohashi, Y., Shimomura, T., and Matsuoka, M., Acquisition of resistance to TMV coincident with induction of pathogenesis-related proteins by TMV infection and chemical treatment in tobacco leaves, *Ann. Phytopathol. Soc. Jpn.*, 52, 626, 1986.
58. Oshima, M., Itoh, H., Matsuoka, M., Murakami, T., and Ohashi, Y., Analysis of stress-induced or salicylic acid-induced expression of the pathogenesis-related 1a protein gene in transgenic tobacco, *Plant Cell*, 2, 95, 1990.
59. Ohashi, Y. and Matsuoka, M., Induction of pathogenesis-related proteins by salicylate or plant hormones in tobacco suspension cultures, *Plant Cell Physiol.*, 28, 573, 1987.
60. Métraux, J. P., AhoGoy, P., Staub, Th., Speich, J., Steinemann, A., Ryals, J., and Ward, E., Induction of systemic resistance in cucumber in response to 2,6-dichloro-isonicotinic acid and pathogens, in *Advances in Molecular Genetics of Plant-Microbe Interaction*, Vol. 1, Hennecke, H. and Verma, D. P., Eds., Kluwer Academic Publishers, Dordrecht, 1991, 432.

CHAPTER 6

Biotechnology for Plant Disease Control

In the broad sense, biotechnology means all technology utilizing biological functions and phenomena to contribute to the welfare of human society. However, the term means, in general, the technology that produces new organisms or analyzes biological function by using gene technology and related techniques.

The breeding of plants resistant against diseases is one of the most effective tools for the control of plant diseases, and many resistant cultivars and varieties of the important crop plants were developed long ago. As the source of resistance genes, wild-types of crop plants have been the target of active search. By hybridization of cultivated plant cultivars with wild-types of plants, the resistance genes are introduced into cultivated plants. However, many undesirable genes in wild plants are also introduced with the resistance genes into cultivated plants by this classic method. To eliminate these undesirable genes from the hybrids, backcrossing of the hybrids to cultivated plants should be done several times. Under these circumstances, it takes a long time to develop one cultivar resistant to one disease. Further, as has been widely known, new races of plant pathogens which attack the newly developed resistant cultivars appear within several years by parasitic adaptation as described.

Recent progress in genetic engineering may shorten the time required for producing new plants, because we can introduce only the isolated useful genes into plants. The other striking usefulness of modern biotechnology is the introduction of genes of different living organisms such as plants, microorganisms, and animals into a living organism; this is in contrast to the classic breeding techniques by which the hybridization, gene transfer, etc. are limited within species, or sometimes, within intergenus.

However, the introduction of a single dominant gene for resistance or a vertical resistance gene into the cultivated plant by gene technology should be avoided, since we may exhaust rapidly gene sources for resistance from this plant because the appearance of new races that overcome new resistance genes occurs by one-step mutation of the pathogens.

Thus, the development of new disease-resistant cultivars by new gene technology should be limited to using horizontal resistance genes or genes for another type of resistance such as detoxification of pathotoxins, or to inducing cross protection.

I. BASIC TECHNIQUES FOR GENETIC RECOMBINATION

The purpose of genetic recombination is to produce a recombinant gene in vitro and to express it after the introduction into plant cells. In general, preparation of the recombinant deoxyribonucleic acid (DNA) and production of a transgenic plant are done by the methods indicated in Figure 1 and by the following procedures.

A. Preparation of Passenger DNA

The passenger DNA can be obtained by digestion of the extracted DNA with restriction enzymes. The other method is to obtain cDNA by reverse transcription of messenger ribonucleic acid (mRNA). Passenger DNA is sometimes synthesized chemically from nucleotides deduced by the amino acid sequence of proteins.

B. Preparation and Construction of Vector DNA (vDNA)

In general, virus DNAs, bacterial plasmids, and phage DNAs are used as the source of vector DNA. Vector DNAs are usually constructed from these DNAs by eliminating unnecessary and undesirable genes, and then introducing promoter and terminator sequences with several marker genes such as tolerant genes for antibiotics or auxotrophic genes.

C. Introduction of Genes into Vector DNA

At first, vector DNA is cut by an appropriate restriction enzyme; for the fundamental functions of DNAs such as replication, marker genes are well preserved. Then the passenger DNAs are inserted and both terminals are connected with DNA lygase. Thus the circular recombinant DNAs are constructed.

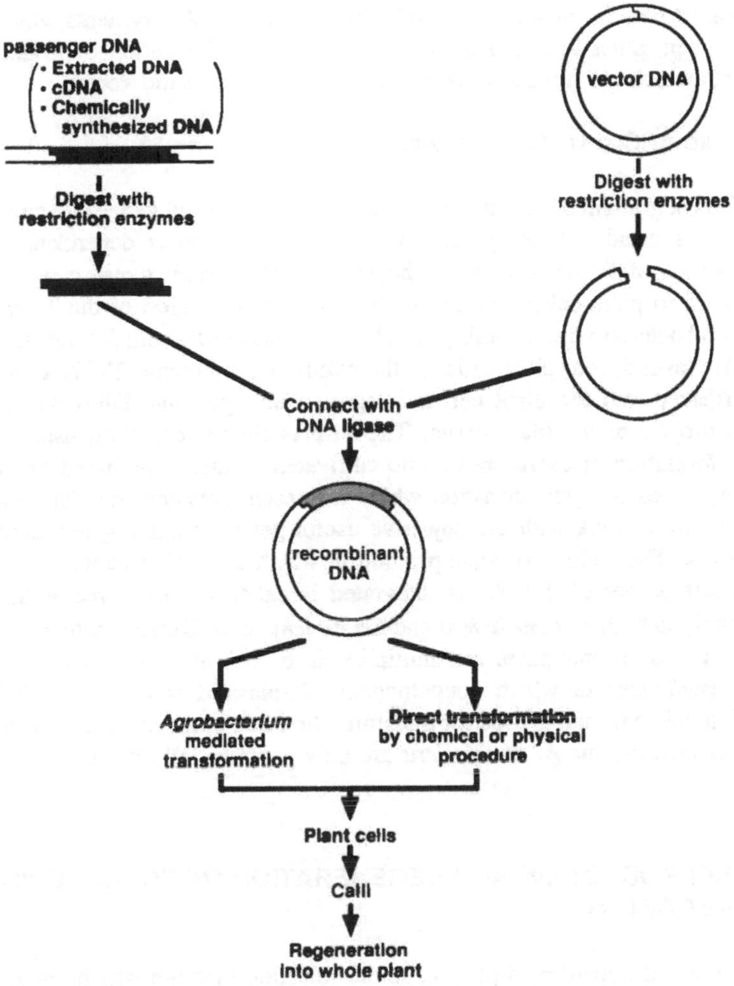

FIGURE 1. Schematic illustration of introduction of a gene into higher plants.

II. TRANSFORMATION OF RECOMBINANT DNA

The transformation of recombinant DNA is done in two ways.

A. Direct Transformation

There are two procedures of direct transformation. One is the chemical method; that is, the recombinant DNAs are transformed into plant protoplast under the presence of calcium phosphate or polyethylene glycol (PEG). The other is so-called electroporation which gives high frequency in the water suspension of protoplast with recombinant DNA.

Direct transformation is also able to transform DNA fragments which contain alien genes and is usually used for the transformation of gramineous plants which are not transformed by using the Ti-plasmid vector.

B. Use of Crown Gall Bacteria and the Ti-Plasmid

Except gramineous plants, the transformation of useful genes into cultivated plants is usually done by this method. This method is dependent on the properties of the crown gall pathogen, *Agrobacterium tumefaciens*, which attaches to plant cells; and on the fact that the *vir*-region of the Ti-plasmid of this bacterium has an ability to splice and transfer so-called T-DNA, a part of Ti-plasmid, into plant cells. In the natural environment, T-DNA which is transferred into the plant cell is integrated into genomic DNA and causes hypertrophy of the plant tissue. The Ti-plasmid vector, often used for the transformation of useful genes into cultivated plants, is prepared by substituting genes for gall formation which is present between the right and left border of T-DNA with the objective useful genes containing regulation sequences. The wide host range plasmid in which the region between the right and left border of T-DNA is integrated is called a binary vector that can multiply in both *A. tumefaciens* and *Escherichia coli*. Binary vectors in which alien genes are integrated are multiplied in *E. coli* and are transformed into *A. tumefaciens* in which nonpathogenic Ti-plasmid is contained (T-DNA region deleted), and is used for transformation into plant cells. Binary vectors and nonpathogenic *A. tumefaciens* are now commercially available.

III. REPRODUCTION AND REGENERATION OF TRANSGENIC PLANT CELLS

To obtain transformed plant cells the selection medium which contains an antibiotic or an auxotrophic medium is usually used, because as described, the plasmid vectors are constructed to have such marker genes.

The selected transformant cells are cultured on the artificial culture medium to grow into the callus. Under the appropriate ratio of indole acetic acid (IAA) and cytokinin, several plant species can be regenerated into whole plants from the calli. Thus, the tissue culture technique is necessary in obtaining the transgenic plants.

IV. REPRODUCTION OF TRANSGENIC PLANTS RESISTANT TO VIRUS DISEASES

The chemical control of a virus disease is usually unsuccessful, and several biocontrol measures have been developed as described in Chapter 5. Further, according to the recent progress in gene technology, many new trials have been conducted to produce virus-resistant plants.

As Zaitlin and Hull[1] pointed out, the resistance of plants against virus diseases is considered to operate at the following three levels:

1. In an extreme case of resistance, termed total immunity, no virus replication takes place, even in the cells initially entered.
2. Some resistant hosts may allow a virus to replicate in initially infected cells, but not to spread from cell to cell.
3. In other hosts some initial replication and spread of a virus may induce a hypersensitive response which restricts the virus to a region around the point of entry.

These three levels are useful in evaluating and analyzing the mode of action of the resistance of transgenic plants.

The three main approaches to producing virus-resistant transgenic plants are classified by the introduction and expression of: (1) coat protein gene of viruses, (2) antisense RNA genes, and (3) satellite RNA genes of viruses.

A. Transgenic Plants With Virus Coat Protein Genes

The coding sequences of coat protein genes of the tomato mosaic virus (TMV) with the 35S promoter of cauliflower mosaic virus (CMV) has been introduced into tobacco and tomato plants, and the transgenic plants were found to be partly resistant to infection by TMV.[2-6] Resistance was expressed as the failure or delay of symptom appearance, and the reduction of virus multiplication. The degree of resistance has been reported to correlate with the amount of coat protein synthesized in the transgenic plants. The protection was effective to many TMV isolates though the degree of protection varied. For example, the transgenic tomato plants which express a coat protein gene of the common (U1) strain of TMV did not develop symptoms or cause virus multiplication of the same virus strain for 30 days after inoculation. However, about 50% of the transgenic plants developed symptoms when inoculated with a more aggressive strain (PV-230), though the appearance of symptoms was delayed.

Motoyoshi[6] introduced the coat protein gene of TMV (tomato strain) into the tomato plant which has the TMV resistance gene, *Tm-2*. This transgenic tomato showed resistance to the TMV mutants which overcame the action of *Tm-2*. This fact shows that the combined use of original resistance genes and coat protein genes may be a very effective measure to the appearance of new virulent virus mutants. He also found that the resistance of the transgenic plant was maintained to the next generation.

Transgenic plants that express coat protein genes in a wide range of viruses such as AlMV,[7] CMV,[8] PVX,[9,10] and so on, also were found to have protective effects against respective viruses. Lawson et al.[11] produced potatoes resistant against both potato virus X (PVX) and potato virus Y (PVY) by introducing coat protein genes of both viruses.

The coat protein expressing tomato plants were also resistant to infection by the tomato strain of TMV in the field, and no yield reduction of fruits was observed in contrast to the control plants which decreased yield reduction 25–30% by infection.[4] These facts indicate that this type of approach is appreciable to produce resistant plants against virus diseases, at least to positive-strand RNA viruses.

Environmental conditions affect the resistance of coat protein expressing transgenic plants. For example, expression of a coat protein and resistance to TMV in transgenic tobacco plants decreased by continuous exposure to high temperature (35°C).[5]

The mechanism of coat protein mediated protection is not clear. However, in some features, the protection is similar to that of the cross protection in which inoculation of the attenuated virus strain results in protection from the virulent strain.[12]

As Register et al.[13] suggested, in these transgenic plants, protection functions at two levels: at the initial site of the infection, and as a block to the long-distance spread. Little or no delay of virus spread was observed in the cell-to-cell spread.[13] Regarding the long-distance spread, a delay in TMV movement into upper leaves was observed in coat protein transgenic tobacco as compared to controls.[2,14]

Studies using protoplasts prepared from transgenic plants with the TMV coat protein revealed that there is an inhibition at an early stage of infection.[15] In protoplasts as well as in whole plants, transgenic plants with the TMV coat protein are protective against the TMV but not against the TMV RNA. The partially uncoated virions (ca. 20 coat proteins, i.e., one turn of the helix) from the 5′ end of the virus overcame protection in the transgenic protoplast (and in the plant).[16] These facts suggest that the protection mechanism operates before the release of viral RNA (uncoating of virus particles).

Stark et al.[16] found that a greater degree of coat protein-mediated protection resulted when polymerized forms of the TMV coat protein were introduced into protoplasts than when free capsid proteins were introduced prior to infection by the TMV. That is, protection seems to be dependent on the form of the coat protein as well as its sequences.

B. Transgenic Plants of Antisense RNA

Antisense RNA has been known to play an important role in the regulation of gene expression of prokaryotes and has been used to suppress gene expression in both prokaryotes and eukaryotes[17,18] including plants. Using this phenomenon, several trials to produce transgenic plants of genes for antisense RNA of the viral RNA, which are resistant to virus diseases have been conducted. In practice, a region of 5′ upstream of the cDNA of the virus

RNA is connected with a promoter and introduced into plant cells. Then the synthesized antisense RNA produces a double-stranded RNA with the virus genome RNA, and inhibits the translation.

As to the mechanism of inhibition of the translation, several hypotheses are available. One is that the transfer of double-stranded RNA to the endoplasmic reticulum may be inhibited. Another one is that the replication of RNA is inhibited by binding the antisense RNA to the binding site of replicase of viral RNA. The third hypothesis is that there is a deficiency of nucleotides in synthesizing viral RNA by competition with antisense RNA. Further, there is a hypothesis that the newly formed double-stranded RNA may be broken down very rapidly.

Transgenic plants expressing antisense RNA to the 3′ region (including coat protein gene) of TMV,[19] PVX,[9] or CMV[8] have some protective effect against infection with respective viruses or RNA. However, the protective effect is not sufficient as compared to transgenic plants expressing the coat protein.

The low level of protection afforded by antisense RNA seems to be due to the low level of suppression of gene expression. According to Robert et al.[18] the suppression of antisense RNA in transgenic plants results in a reduction of 90–95% in gene expression, and this degree of reduction still allows virus multiplication; hence, the protective effect seems to be insufficient.

The efficiency of gene expression is increased by increasing promoter activity, for example, by introducing additional enhancer elements or by increasing the number of template copies. Amplification vectors of the gemini virus gene are considered to be suitable for the latter purpose since a master copy can be stably integrated into the plant genome, and gene expression is independent of the position effect.[20]

In plant viruses, the effect of antisense RNA to virus disease control has been tested only in RNA viruses; therefore, the protection may be limited to cytoplasmic effects of antisense RNA. Buck[21] considered that antisense RNA may more effectively control DNA viruses, such as gemini viruses, which replicate in the nucleus and therefore would be susceptible to the suppressive effects of antisense RNA in both the nucleus and cytoplasm.

C. Transgenic Plants Expressing Satellite RNAs

1. Satellite Virus and Satellite RNA

There are viruses which do not multiply by themselves according to the lack of genetic information. These viruses are called satellite viruses. The first discovery in plant satellite viruses is the tobacco necrosis satellite virus (TNSV) which became reproducible only under the coexistence of the tobacco necrosis virus (TNV). In this case the TNV is called a helper virus. That is, the multiplication of TNSV is completely dependent on the TNV genome. The TNSV is composed of 1239 nucleotides, and only the coat protein gene is located in this sequence.

Some viruses such as the tobacco ring spot virus (TRSV), cucumber mosaic virus (CMV), etc. were found to have an extra small RNA, in addition to the genomic RNA; and they were termed as satellite RNAs. Satellite RNAs are also unable to multiply by themselves. Satellite RNAs are totally dependent on helper viruses for their multiplication, and they do not code coat proteins. That is, the coat protein synthesis is helper virus dependent. Satellite RNAs are usually compound of 334–386 nucleotides.

The association between a satellite RNA and its helper virus is specific; that is, related viruses can act as helpers for a given satellite. The satellite RNAs are not required for the replication of their helper viruses.

2. Trials to Prevent the Multiplication of Helper Viruses

Satellite viruses, satellite RNAs, and defective-interfering RNAs have an ability to interfere in the multiplication of helper viruses. Because of the property of interference, attempts have been made to introduce genes for satellite viruses, satellite RNA, and defective interfering RNA into plants to control virus diseases caused by helper viruses by expressing them within the transgenic plants. It may be theoretically possible that genes for these small RNAs can be synthesized artificially, integrated into plant genomes, and expressed from these plants.

The replication of satellite RNAs and satellite viruses in transgenic plants occurs only after infection by helper viruses. Gerlach et al.[22] produced transgenic tobacco plants which have integrated trimers of the cloned permuted monomers of (+) and (−) strands of the tobacco ring spot satellite virus cDNA with a 35S promoter and *nos* terminator. Both transgenic plants contained trimers in orientation which allowed transcription of (+) and (−) satellite RNAs, respectively. The major transcripts from (+) and (−) transgenic plants were of monomer length. This was proved by Northern blot analyses, which means the autocatalytic cleavage of transcripts occurred.

When transgenic plants were infected by TRSV, there was a large increase in the level of satellite RNA sequences. The striking feature of satellite (+) RNA transgenic plants infected by TRSV was the attenuation of symptoms as compared with control plants. Six days after inoculation with TRSV, the growth of transgenic plants was similar to that of uninfected plants, when the control plants were stunted and newly developed leaves showed severe mottling.

In transgenic plants of satellite (+) RNA, the attenuated symptoms correlated with a lower level of virus replication in upper symptomless leaves. Transgenic plants of satellite (−) stranded RNA appear to be susceptible in the initial stage after infection. However, plants developed resistance after 6 weeks.

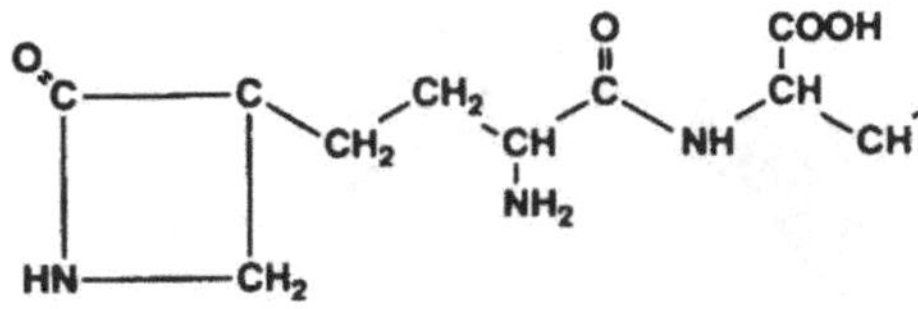

Tabtoxin

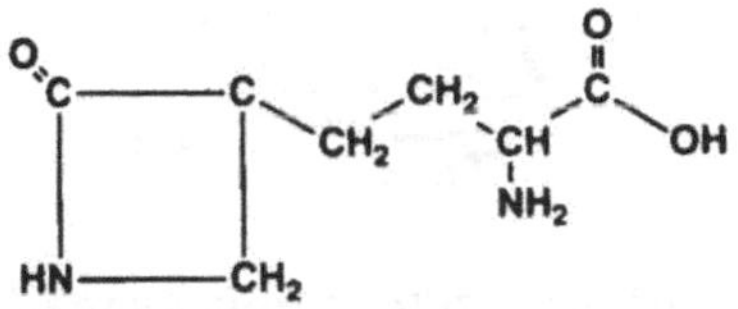

Tabtoxinine-β-lactam

FIGURE 2. Tabtoxin and the metabolite tabtoxinine-β-lactam.

Transgenic tobacco plants integrated with satellite RNA of the CMV are reported to be resistant to CMV as well.[23–25] Transgenic plants containing DNA of satellite RNA with a 35S promoter were obtained. When these transgenic plants were challenged with CMV, symptoms were greatly attenuated and the growth of plants was almost the same as uninoculated plants. The attenuation of symptoms correlated with the amount of satellite RNA formed and is considered to be due to the inhibition of virus multiplication.

The resistance of transgenic plants expressing CMV satellite RNA is effective against the helper virus and closely related viruses, but not against the infection by viruses of different taxonomic groups.

Good protection is obtained with these genetically engineered satellite RNAs; and unlike the coat protein-mediated protection, the protective effect is not overcome at a high inoculum level or with time. In spite of these results, there have been reports that the satellite RNA mutated in several nucleotides causes more severe diseases.[26,27]

V. PRODUCTION OF TRANSGENIC PLANTS RESISTANT TO BACTERIAL DISEASES

The wildfire pathogen of tobacco *Pseudomonas syringae* pv. *tabaci* has been known to produce a toxin which causes the chlorotic symptom. The toxin, dipeptide, was named tabtoxin (Figure 2). It is indicated that in plants some peptidases convert tabtoxin to tabtoxinine-β-lactam,[28] which inhibits

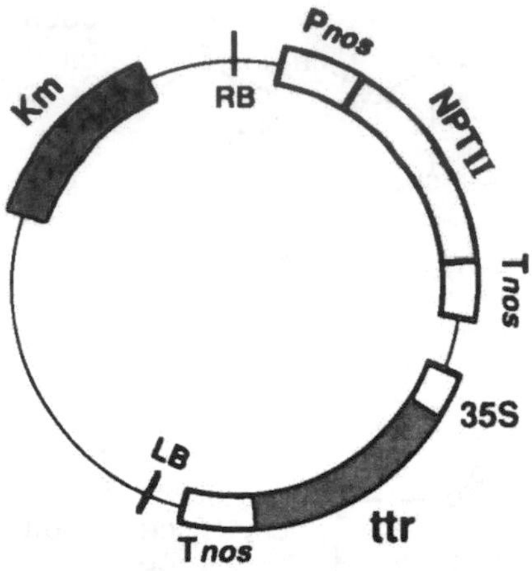

FIGURE 3. Plant vector pARK21 having tabtoxin detoxification gene *ttr*. Km: Kanamycin-resistant gene; P*nos:* promoter of nopaline synthase gene; T*nos:* terminator of nopaline synthase gene; NPTII: neomycin phosphotransferase II gene; RB and LB: right border and left border sequences of T-DNA; 35S: cauliflower mosaic virus 35S promoter. (From Anzai, H., Yoneyama, K., and Yamaguchi, I., *Mol. Gen. Genet.*, 219, 492, 1989. With permission.)

glutamine synthase;[29] and as a result, abnormal accumulation in tobacco cells of ammonium causes characteristic chlorosis.

Tabtoxin, a nonspecific toxin, is toxic to a wide range of living beings including bacteria. However, the tabtoxin-producing bacterium *P. syringae* pv. *tabaci* is tolerant to this toxin, suggesting that it produces the detoxification enzyme for tabtoxin. Anzai et al.[30] and Yoneyama and Anzai[31] isolated the gene encoding enzyme, acetyltransferase *(ttr)*, which detoxifies tabtoxin from the genomic DNA of wildfire bacterium. This gene was fused to a 35S promoter to construct a chimeric gene and was introduced into tobacco cells by *Agrobacterium*-mediated transformation (Figure 3). Transgenic tobacco plants showed a high specific expression of *ttr* gene and did not cause chlorosis by treatment with tabtoxin. They were also resistant to the wildfire bacterium. These results indicate not only that tabtoxin plays a major role in pathogenesis in the wildfire disease but also that a disease-resistant plant can be produced by introducing the toxin-detoxification gene into plants. The nucleotide sequence of *ttr* gene was determined. The open reading frame (ORF) of 531 base pairs (bp) encodes a protein composed of 177 amino acids.[32]

Düring et al.[33] introduced the bacteriophage T4 lysozyme gene into potato plants. Transgenic plants have been shown to provide a significantly reduced susceptibility to infection by *Erwinia carotovora*. In vitro assays demonstrate a much smaller extent of tuber tissue maceration. Sprouting assays for infected tissue showed a higher resistance to bacterial infection in comparison to control tissue. They concluded that the secretion of a bacteriolytic enzyme into the intercellular space may be the key factor in inhibiting the invading bacteria.

VI. PRODUCTION OF TRANSGENIC PLANTS RESISTANT TO FUNGAL PATHOGENS

Many attempts to produce transgenic plants resistant to fungal pathogens have been conducted. Introduction of the chitinase gene, glucanase gene, protease inhibitor gene, and key genes involved in phytoalexin biosynthesis into plants seems to be a promising strategy for the control of fungal diseases.

A. Transgenic Plants Expressing Chitinase

Chitinase is known to be an antifungal enzyme that degrades chitin, a main component of fungal cell walls except in fungi belonging to Phycomycetes. The role of chitinase in the defense response of plants has largely been dependent on data obtained from in vitro studies. The transgenic plants that express foreign chitinase genes may throw a light on the actual role of chitinase in the defense reaction of plants against fungal pathogens.

The gene sources for transgenic plants are microorganisms and plants. Jones et al.[34] constructed a bacterial chitinase gene from *Serratia marcescens (ChiA)* fused to a promoter of the ribulose bisphosphate carboxylase subunit *(rbc S1)* gene, and promoters of two different chlorophyll a/b binding protein *(cab)* genes from the petunia. These structures were introduced to a plant vector, derived from the *Agrobacterium* Ti-plasmid, and used to generate transgenic tobacco plants; and the amount of *ChiA* mRNA was compared. As the result, *rbcS/ChiA* fusion gave rise to threefold more *ChiA* mRNA than *cab/ChiA* fusion. The chitinase was expressed at a higher level in *rbcS/ChiA* fusion than in *cab/ChiA* fusion and control plants.

Chitinase genes originating from plants were also introduced into plants in order to produce resistant plants against fungal pathogens. Neuhaus et al.[35] transformed *Nicotiana sylvestris* cells with sequences encoding a class I (basic) chitinase of tobacco regulated by cauliflower mosaic virus 35S expression signals. The gene was expressed to give mature, enzymatically active chitinase to the intracellular spaces of leaves. Most transformants accumulated an extremely high level of chitinase, for example, 120-fold that of nontransformed plant tissues.

In this case, they found that a high chitinase level in a transgenic plant did not increase the resistance against *Cercospora nicotiana*, a chitin-containing fungus. In other words, basic chitinase does not appear to be the factor of defense reaction to this pathogen.

Broglie et al.[36] constructed a hybrid chitinase gene by replacement of the 5′ regulatory region of the bean endochitinase gene *(CH5B)* with the promoter region of cauliflower mosaic virus 35S transcript. The chimeric 35S chitinase genes are inserted into plasmid vectors, mobilized from *Escherichia coli* into *Agrobacterium tumefaciens*, and allowed to infect leaf disks of *Nicotiana tabacum* cv. *Xanthi*. They obtained transformants that express bean chitinase, which is ascertained by immunoblot analysis. The expressed bean chitinase proteins were somewhat higher in amount in the root of transgenic tobacco plants than in the leaves.

These transgenic tobacco plants were assayed for resistance to *Rhizoctonia solani*, a soilborne fungal pathogen. As a result, transgenic plants showed greater resistance to the development of disease symptoms. When 35S chitinase transgenic plants were grown in the presence of *Pythium aphanidermatum*, which lacks chitin in the cell wall, no difference in survival was detected compared to control plants. Thus, the transgenic tobacco plant expressing a bean chitinase gene under the control of a 35S promoter showed a protective effect against *Rhizoctonia solani*, a chitin-containing fungus.

Nishizawa and Hibi[37] isolated the rice chitinase cDNA clone, and found that rice chitinase showed 67% homology to the basic chitinases from tobacco and beans, but the codon usage was different from chitinases of dicot plants.

Nishizawa et al.[38] further introduced rice chitinase cDNA with the 35S promoter into binary vector pBI 121, transferred to *Agrobacterium;* and the tobacco plant was transformed with the rice chitinase gene. The transgenic tobacco plants were highly resistant to *Erysiphe cichoracearum*, the powdery mildew fungus, showing strong inhibition of colony development.

B. Transgenic Plants Expressing Barley Ribosome-Inactivating Protein

Leach et al.[39] isolated cDNA from barley seeds encoding a single chain, type I ribosome-inactivating protein (RIP). RIP does not inactivate "self" ribosomes, but inactivates in varying degree toward ribosomes of distantly related species. Purified barley RIP inhibits the growth of fungi in vitro, and the inhibition is synergistically enhanced in the presence of fungal cell wall degrading enzymes.

Logemann et al.[40] obtained transgenic tobacco plants that express barley RIP, in order to know whether this gene is a useful candidate for plant protection against fungal pathogens. In practice, the chimeric gene containing a transcriptional fusion between barley RIP cDNA and the promoter of the wound- and pathogen-inducible potato *wun1* gene was constructed. The chimeric gene was cloned into the binary vector bin 19 and transformed into *Agrobacterium tumefaciens* strain LBA4404, and tobacco plants cv. SR I was transformed.

These transformants were selected on kanamycin, and tested for growth in soil inoculated with *Rhizoctonia solani*. The transgenic tobacco plants grew well compared to control plants in the infested soil. They further ascertained that the transgene is inherited and expresses barley RIP upon wounding.

VII. OTHER MEANS OF ENGINEERING DISEASE RESISTANCE

The inhibition of essential enzymes of pathogens to evoke disease is a useful tool for disease control.[41] The transgenic plants expressing such inhibitor genes might protect plants from pathogens.

The genes encoding antimicrobial peptides seem to be effective candidates for producing disease resistant plants, and such genes have been described from plants themselves[42–44] and also from animals.[45–48]

Antibody genes encoding immunoglobulins have been reported to be expressed in plants.[49] This finding suggests the possibility that transgenic plants with genes encoding antibodies to important enzymes such as virus replicase, fungal suppressors, toxins, and cell wall degrading enzymes may directly inactivate the pathogenicities of pathogens.[50]

REFERENCES

1. Zaitlin, M. and Hull, R., Plant virus-host interactions, *Annu. Rev. Plant Physiol.*, 38, 291, 1987.
2. Powell, P. A., Nelson, R. S., De, B. N., Hoffmann, N., Rogers, S. G., Fraly, R. T., and Beachy, R. N., Delay of disease development in transgenic plants that express the tobacco mosaic virus coat protein gene, *Science*, 232, 738, 1986.
3. Nelson, R. S., Powell, A. P., and Beachy, R. N., Lesion and virus accumulation in inoculated transgenic tobacco plants expressing the coat protein gene of tobacco mosaic virus, *Virology*, 158, 126, 1987.
4. Nelson, R. S., McCormick, S. M., Delannay, X., Dube, P., Layton, J., Anderson, E. J., Kaniewska, M., Proksch, R. K., Horsch, R. B., Roers, S. G., Fraley, R. T., and Beachy, R. N., Virus tolerance, plant growth, and field performance of transgenic tomato plants expressing coat protein from tobacco mosaic virus, *Biotechnology*, 6, 403, 1988.
5. Nejidat, A. and Beachy, R. N., Decreased level of TMV coat protein in transgenic tobacco plants at elevated temperatures reduce resistance to TMV infection, *Virology*, 173, 531, 1989.
6. Motoyoshi, F., Virus resistant transgenic plants with virus coat protein genes, *Tissue Cult.*, 16, 13, 1990.
7. Loesch-Fries, L. S., Merlo, D., Zinnen, T., Burhop, L., Hill, K., Kraha, K., Vavis, N., Nelson, S., and Halk, E., Expression of alfalfa mosaic virus RNA 4 in transgenic plants confers virus resistance, *EMBO J.*, 6, 1845, 1987.
8. Cuozzo, M., O'Connell, K. M., Kaniewski, W., Fang, R.-X., and Turner, N. E., Viral protection in transgenic tobacco plants expressing the cucumber virus coat protein or its antisense RNA, *Biotechnology*, 6, 549, 1988.
9. Hemenway, C., Fang, R.-X., Kaniewski, W. K., Chua, N.-H., and Turner, N. E., Analysis of the mechanism of protection in transgenic plants expressing the potato virus X coat protein or its antisense RNA, *EMBO J.*, 7, 1273, 1988.
10. Hoekema, A., Fusiman, M. J., Molendijk, L., Van den Elzen, P. J. M., and Cornelissen, B. J. C., The genetic engineering of two commercial potato cultivars for resistance to potato virus X, *Biotechnology*, 7, 273, 1989.
11. Lawson, C., Kaniewski, W., Haley, L., Rozman, R., Newell, C., Sanders, P., and Turner, N. E., Engineering resistance to mixed virus infection in a commercial potato cultivar: resistance to potato virus X and potato virus Y in transgenic Russet Burbank, *Biotechnology*, 8, 127, 1990.

12. Ponz, F. and Bruening, G., Mechanism of resistance to plant viruses, *Annu. Rev. Phytopathol.*, 24, 355, 1986.
13. Register, J. C., III, Powell, P. A., Nelson, R. S., and Beachy, R. N., Genetically engineered cross protection against TMV interferes with initial infection and long-distance spread of the viruses, in *Molecular Biology of Plant-Pathogen Interactions*, Staskawicz, B., Alquist, P., and Yoder, O., Eds., Alan R. Liss, New York, 1988, 269.
14. Anderson, E. J., Stark, D. M., Nelson, R. S., Powell, P., Tumer, N. E., and Beachy, R. N., Transgenic plants that express the coat protein genes of TMV or AlMV interfere with disease development of some nonrelated viruses, *Phytopathology*, 79, 1284, 1989.
15. Register, J. C., III and Beachy, R. N., Resistance to TMV in transgenic plants results from interference with an early event in infection, *Virology*, 166, 524, 1988.
16. Stark, D. M., Register, J. C., III, Nejidat, A., and Beachy, R. N., Toward a better understanding of coat protein mediated protection, in *Plant Gene Transfer*, Lamb, C. J. and Beachy, R. N., Eds., Alan R. Liss, New York, 1990, 275.
17. Green, P. J., Pines, O., and Inouye, M., The role of antisense RNA in gene regulation, *Annu. Rev. Biochem.*, 55, 569, 1986.
18. Robert, L. S., Donaldson, P. A., Ladaique, C., Altossar, I., Arnison, P. G., and Fabijanski, S. F., Antisense RNA inhibition of β-glucuronidase gene expression in transgenic tobacco plants, *Plant Mol. Biol.*, 13, 399, 1989.
19. Powell, P. A., Stark, D. M., Sanders, P. R., and Beachy, R. N., Protection against tobacco mosaic virus in transgenic plants that express tobacco mosaic virus antisense RNA, *Proc. Natl. Acad. Sci. U.S.A.*, 86, 6949, 1989.
20. Hayes, R. J., Coutts, R. H. A., and Buck, K. W., Stability and expression of bacterial genes in replicating geminivirus vectors in plants, *Nucl. Acids Res.*, 17, 2391, 1989.
21. Buck, K. W., Virus resistant plants, in *Plant Genetic Engineering*, Vol. 1, Grierson, D., Biol, C., and Biol, F. I., Eds., Blackie, Glasgow, 1990, 136.
22. Gerlach, W. S., Llewellyn, D., and Haseloff, J., Construction of a plant disease resistance gene from the satellite RNA of tobacco ring spot virus, *Nature (London)*, 328, 802, 1987.
23. Baulcombe, D. C., Saunders, G. R., Bevan, M. W., Mayo, M. A., and Harrison, B. D., Expression of a biologically active viral satellite RNA from the nuclear genome of transformed plants, *Nature (London)*, 321, 446, 1986.
24. Harrison, B. D., Mayo, M. A., and Baulcombe, D. C., Virus resistance in transgenic plants that express cucumber mosaic satellite RNA, *Nature (London)*, 328, 799, 1987.
25. Palukaitis, P., Pathogenicity regulation by satellite RNAs of cucumber mosaic virus: minor nucleotide sequence changes alter host responses, *Mol. Plant-Microbe Interact.*, 4, 175, 1988.
26. Devic, M., Jaegle, M., and Baulcombe, D., Symptom production on tobacco and tomato is determined by two distinct domains of the satellite RNA of cucumber mosaic virus (strain Y), *J. Gen. Virol.*, 70, 2765, 1989.
27. Kurath, G. and Palukaitis, P., Satellite RNAs of cucumber mosaic virus: recombinants constructed in vitro reveal independent functional domains for chlorosis and necrosis in tomato, *Mol. Plant-Microbe Interact.*, 2, 91, 1989.

28. Uchytil, T. F. and Durbin, R. D., Hydrolysis of tabtoxins by plant and bacterial enzymes, *Experimentia*, 36, 301, 1980.
29. Sinden, S. L. and Durbin, R. D., Glutamine synthetase inhibition: possible mode of action of wildfire toxin from *Pseudomonas tabaci*, *Nature (London)*, 219, 379, 1968.
30. Anzai, H., Yoneyama, K., and Yamaguchi, I., Transgenic tobacco resistant to a bacterial disease by the detoxification of a pathogenic toxin, *Mol. Gen. Genet.*, 219, 492, 1989.
31. Yoneyama, K. and Anzai, H., Gene technological study on disease control by the inactivation of pathogenic toxins in transgenic plants, *J. Pest. Sci.*, 16, 291, 1991.
32. Anzai, H., Yoneyama, K., and Yamaguchi, I., The nucleotide sequence of tabtoxin resistance gene *(ttr)* of *Pseudomonas syringae* pv. *tabaci*, *Nucleic Acids Res.*, 18, 1890, 1990.
33. Düring, K., Fladung, M., and Lörz, H., Antibacterial resistance of transgenic potato plants producing T4 lysozyme, *Program & Abstr., 2nd EFPP Conf.*, Strasbourg, 1992, 135.
34. Jones, J. D. G., Dean, C., Gidoni, D., Gilbert, D., Nutter, D., Lee, R., Bedbrook, J., and Dunsmuir, P., Expression of bacterial chitinase protein in tobacco leaves using two photosynthetic gene promoters, *Mol. Gen. Genet.*, 212, 536, 1988.
35. Neuhaus, J., Ahl-Goy, P., Hinz, U., Flores, S., and Meins, F., High-level expression of a tobacco chitinase gene in *Nicotiana sylvestris*. Susceptibility of transgenic plants to *Cercospora nicotianae* infection, *Plant Mol. Biol.*, 16, 141, 1991.
36. Broglie, K., Chet, I., Holliday, M., Gressman, R., Diddle, P., Knowlton, S., Mauvais, C. J., and Broglie, R., Transgenic plants with enhanced resistance to the fungal pathogen *Rhizoctonia solani*, *Science*, 254, 1194, 1991.
37. Nishizawa, Y. and Hibi, T., Rice chitinase gene: cDNA cloning and stress-induced expression, *Plant Sci.*, 76, 211, 1991.
38. Nishizawa, Y., Kondo, K., Akutsu, K., and Hibi, T., Disease-resistant transgenic tobacco expressing a rice chitinase gene, *Abstr. Pap. Annu. Meet. Phytopathol. Soc. Jpn.*, 1992, 45.
39. Leach, R., Tommerup, H., Svendesen, I., and Mundy, J., Biochemical and molecular characterization of three antifungal proteins from barley seed, *J. Biol. Chem.*, 266, 1564, 1991.
40. Logemann, J., Jach, G., Tommerup, H., Mundy, J., and Schell, J., Expression of a barley ribosome-inactivating protein leads to increased fungal protection in transgenic tobacco plants, *Biotechnology*, 10, 305, 1992.
41. Cervone, F., Delorenzo, G., Degra, L., Salvi, G., and Bergami, M., Purification and characterization of a polygalacturonase-inhibiting protein from *Phaseolus vulgaris* L, *Plant Physiol.*, 85, 631, 1987.
42. Broekert, W., Lee, H.-I., Kush, A., Chua, N.-H., and Raikhel, N., Wound-induced accumulation of mRNA containing a hevein sequence in laticifers of rubber tree *(Hevea brasiliensis)*, *Proc. Natl. Acad. Sci. U.S.A.*, 87, 7633, 1990.
43. Chrispeels, M. J. and Raikhel, N. V., Lectins, lectin genes, and their role in plant defense, *Plant Cell*, 3, 1, 1991.

44. Vigers, A. J., Roberts, W. K., and Selitrennikoff, C. P., A new family of plant antifungal proteins, *Mol. Plant-Microbe Interact.*, 4, 315, 1991.
45. Boman, H. G. and Hultmark, D., Cell-free immunity in insects, *Annu. Rev. Microbiol.*, 41, 103, 1987.
46. Dimarcq, J.-L., Zachary, D., Hoffmann, J. A., Hoffmann, D., and Reichart, J.-M., Insect immunity: expression of the two major inducible antibacterial peptides, defensin and diptericin, in *Phormia terranovae*, *EMBO J.*, 9, 2507, 1990.
47. Lehrer, R. I., Ganz, T., and Selsted, M. E., Defensins: endogenous antibiotic peptides of animal cells, *Cell*, 64, 229, 1991.
48. Zasloff, M., Magainins, a class of antimicrobial peptides from *Xenopus* skin: isolation, characterization of two active forms, and partial cDNA sequence of a precursor, *Proc. Natl. Acad. Sci. U.S.A.*, 84, 5449, 1987.
49. Hiatt, A., Cafferkey, R., and Bowdish, K., Production of antibodies in transgenic plants, *Nature (London)*, 342, 76, 1989.
50. Keen, N. T., The molecular biology of disease resistance, *Plant Mol. Biol.*, 19, 109, 1992.

INDEX

A

F

I

Q

R

U

V

W

X

Y

Milton Keynes UK
Ingram Content Group UK Ltd.
UKHW040057071024
449327UK00019B/621

9 780367 449698